The Essential Guide to Special Education in Ireland

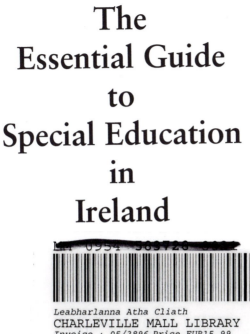

Dr. David J. Carey

Primary /

D0320983

PUBLISHED BY:
PrimaryABC
Abercorn House
57 Charleston Road
Ranelagh
Dublin 6
Phone: (01) 269 5008
Email: primaryabc@eircom.net
Website: www.primaryabc.ie

Typesetting: Keystrokes Digital, Dublin

Printed and bound by: ColourBooks Ltd, Dublin

ISBN: 0-9545837-2-8

Contents

About the author

Dr. David J. Carey is the Co-ordinator of Special Education and Programme Development at the Froebel College of Education, Blackrock, Co. Dublin, Ireland. A psychologist with twenty-eight years of experience in educational and clinical settings, Dr. Carey lectures at the College to undergraduate and post-graduate students. He is the founding editor of the Froebel Journal of Child-Centred Education and sits on the Editorial Board of REACH, the Journal of the Irish Association of Teachers in Special Education.

He has provided expert testimony in a large number of High Court cases in Ireland. Dr. Carey is also engaged in teacher training internationally, currently directing a diploma programme for Kindergarten and primary teachers in Kenya and East Africa. He is the author of numerous articles in educational, psychological and lay journals. Dr. Carey lives in Dublin, Ireland and Akrata, Greece.

Acknowledgements

To my mother who saw the good in everyone; to Erin, Bridget, Matthew, Aidan, and to Rena, those I love; to all the families who extended to me the privilege of entering their lives and the lives of their children; to all the families in Ireland who have struggled so hard against overwhelming odds and outright opposition to secure appropriate education for their children; to the children; to all the good teachers I have had the pleasure to work with over the years, especially Sr. Catherine Sheehy who helped me understand how important acceptance is to children. I dedicate this book in the hope it will be of some small help.

I also want to thank my publisher and editor, Brian Gilsenan, whose infinite patience, constant support and encouragement, and analytical skills improved this book and made it possible for me to write it.

2005, David J. Carey

Introduction

This is a book for parents, teachers, students in the Colleges of Education, and anyone interested in understanding special education in Ireland. It is written in simple language, without complicated jargon (in so far as possible), and takes as its primary aim the need for people to have a single resource that outlines all the relevant information about special education in the Irish context. The focus throughout the book is always on children, whatever age they might be. I take the point of view that nothing is more important than children and that the needs of the children supersede the needs of the system and the people employed within the system. It is my uncompromising view that children with special education needs in particular, and all children in general, must be held in the utmost position of importance at all times.

At the time of writing (September 2005), this book contains information that is up to date and accurate. The reader must however keep in mind the rapidly changing pace of special education in Ireland. Change is happening at such a fast pace that even those most in touch with special education in Ireland on a daily basis are finding it hard to keep current. There are new structures, new Departmental guidelines and most importantly of all, new legislation. The impact of these new developments has been profound and the impact of them in the future will be equally profound. It is necessary to be aware that some of the information in the book may become quickly outdated. Every effort has been made to present policies as they currently exist but the reader is strongly advised to track developments on the Internet by accessing the websites identified in the appendix, page 263.

Frustration...

I should offer another word of advice at this point. Parents and teachers are often impatient and frustrated with the state of special education in Ireland and with the pace of change. I share that sense of frustration. However, I think it is especially important for us all to remember that Ireland is in the rather unique position of attempting to create a system of special education in a period of less than ten years that other developed countries have taken twenty to thirty years to create! If one keeps this perspective in mind, the frustration may be easier to bear. Change, in order to be effective, needs to happen slowly and with considerable reflection and insight. Ireland needs further developments in special education provision but it is important to learn from the mistakes of other countries and adapt, not simply adopt their structures, regulations and policies.

This is a most dynamic time for special education in Ireland. Many excellent people are working hard to change the system and make it work. It is important to hasten slowly, cooperatively, without bitterness and to always keep the children foremost in mind, in order to achieve the incredibly important task of improving the educational welfare of children with special needs.

About the book

This book is entitled the essential guide because in the view of the author it contains the information that is absolutely necessary to understand and work within the special education system in Ireland. Any book of this type, which professes to be an essential book, requires choices on the part of the author; choices about what documents to include and exclude; choices about what are the essential elements to review and discuss; choices about the critical elements of special education conditions. It is my hope that I have made the correct choices. My readers will be the ones to decide. To them I affirm that I have made my best effort to include what is truly essential. The book was written with parents, teachers, and students in the Colleges of Education and University Departments of Education all equally in mind. It is organised into various parts for ease of reference:

Part One begins with an introduction to the process of special education in Ireland. It first describes the referral process and includes information on interviewing and observation. It then describes the assessment process in detail. Part One concludes with the vital area of educational planning for children with special education needs.

Part Two provides a detailed description of the nature, characteristics, signs and symptoms of the various special education conditions that are recognised as having an entitlement to specialist education in Ireland.

Part Three contains a variety of articles that will help you to better understand children with special education needs. Topics include 'the essential elements of special education' (mainstreaming, integration etc.); the role of play; managing scarce resources; whole school planning; developing social skills and an important article on understanding challenging behaviour. There then follows several articles specific to adolescents and how they grow and develop; the occurrence of what are sometimes called 'hidden disabilities'; information about how the brain grows and develops during the teenage years. Part Three concludes with an article about the importance of partnership between all of the parties involved in special education. This section of the book includes an overview of what parents want from teachers and what teachers want from parents.

Part Four describes the principle of curriculum differentiation, outlining some of the ways that class and specialist teachers can go about planning for children with special education needs. It is not meant as a reference guide but rather a guide to thinking about special education, from a mainstream perspective. This section presents the general issues that a teacher may consider while creating an inclusive classroom, or when working in collaboration with a specialist teacher who is providing special education to a child.

Part Five deals extensively with the existing legislation. Some of it makes for difficult reading – I have tried to distil legal documents into a comprehensible form but that's not always easy. This section is extremely important however because without adequate knowledge of the structure of special education in Ireland, parents and teachers will be entirely dependent on second-hand information. There is no substitute for wading through the primary sources of information but it is my hope that I have provided an adequate summary of them.

Part Five also contains substantial information regarding the current Department of Education and Science guidelines related to special education in Ireland, describing the various bodies involved and also the individual roles of the key people working in special education

Part Six describes two of the most important court cases to have been heard in recent years regarding special education in Ireland, namely the O'Donoghue case and the Sinnott case

Part Seven provides an overview of two important reports, one from the Autism Task Force (2001) and one from the Dyslexia Task Force. These reports offer significant information and interesting recommendations that are described.

Part Eight is where I get to state my hopes, recommendations and advice on the direction of special education in Ireland in future years.

Language
A word about political correctness; many people in Ireland and elsewhere object to terminology such as 'disabled', 'disability', 'special', etc. I have reservations about these terms myself. However, as they are in common parlance in legislation, policy and everyday usage in schools, I have decided to use them in this book. They do not represent a frame of mind on the part of myself and no offence is meant in their usage.

Finally...
This is not a 'how-to' book. It was not written with the intention of providing easy answers to difficult questions of how to teach, or raise, a child with a special need. The reader can find other volumes that achieve that end. I have written this book in the hope that it will provide useful information about the special education system as it is currently constructed

and evolving in Ireland. Parents and teachers need information and they need it now. Things are changing at a rapid pace. To strive to provide appropriate education for children with special needs, one must begin with an understanding of how the system works. Special education conditions carry with them an entitlement to appropriate educational services. We must open our minds and hearts to the lives of these children; recognise that they bring great richness to the home, classroom and society, a richness that is badly needed in this world of conformity. It helps us to resist the impulse from which we tend to believe that all people must think alike, look alike, act alike and share the same beliefs.

The reader will notice that the book takes a more in-depth look at special education provision in the primary than the secondary sector. The reason for this is because it is within the primary sector that special education provision has advanced most vigorously. The provision of special education in the secondary sector is relatively undeveloped at present. There are no mature programmes to investigate or review, that meet an international best-practice standard. However, virtually all of the information contained in the book relates to both the primary and secondary sector. The reader interested in general issues relating to adolescents with learning difficulties should turn to Part Three for some interesting articles.

There are also sections of the book that include some useful detail about educating the Gifted child, a category of special education entitlement that is under-recognised in practice at the present time.

Prologue

The Essentials of Special Education in Ireland

One of the most common questions people ask, parents and teachers alike, relates to the process of special education as it currently is structured in Ireland. What I propose to do at the outset is explain what happens, from beginning to end, when a child is suspected of having a special education condition or a learning difficulty. Be aware however that things are changing so fast – even at the time of writing this, (5th September 2005), another important new Department of Education and Science Circular related to special education has just been issued.

Let's start by explaining the **difference between a special education condition and a learning difficulty**. You will discover in Part Two that a special education condition is one of a number of specific difficulties that are recognised by the Department of Education and Science. Each condition carries with it an entitlement to support and assistance, either in the mainstream classroom, special class or special school.

A learning difficulty is not necessarily a special education condition. The term 'Learning Difficulty' is used to describe children who have problems in reading, mathematics or other academic subjects. It may also be used to describe a child who is having problems in the areas of physical, emotional, social or language skills. Children with learning difficulties need assistance, support and guidance, but not usually at the level of intensity or duration that children with special education conditions need. Keep this in mind when you are talking to a teacher, and if you are a teacher keep this in mind when talking to a parent. No child can have a special education condition until there has been a complete assessment, either psychological, medical, speech/language, psychiatric, neurological or otherwise. Children with learning difficulties may be assessed by a Learning Support Teacher (see page 195) but until one of the assessments mentioned above has been completed, they should not be referred to as having a special education condition.

If a parent or teacher has concerns that a child has a learning difficulty, a process known as the **'Staged Approach'** is initiated. In the staged approach, there are a number of steps that must be taken by teachers and parents, working together as a team. Let's begin by using an example. David is in first class and his mother has noticed that he does not seem to be able to write any of his letters legibly. Not only are they poorly formed but the spaces between them are either much to large or much too small. The letters extend below and above the lines of the paper. He holds his pencil awkwardly and complains that his hand hurts when he tries to write. David's mother asks his teacher if she has noticed this and the teacher says she has; moreover the teacher believes this pattern is quite unlike that of other boys his age; she has tried different strategies to help David and continues to be concerned.

At this point in the process the teacher and the parent will begin to make a short intervention plan. The teacher will make some careful observations of David when he is writing. She might give him some fine-motor tasks (see glossary) that don't involve writing, to try to ascertain if he has difficulty with these skills. Alternatively the teacher might complete a rating scale (see glossary) that will help determine if David's difficulty is really significantly unusual for a child his age.

Using the information she gathers from observation and informal assessment, the teacher will make up a written plan of things she is going to do with David over the next two school terms. The plan will be shared with David's parents and might include suggestions about how they can help David with his handwriting. This intervention plan will continue for two school terms and from time to time, the teacher will discuss David's progress with the parents and find out from them how they think David is getting on.

If, after about two school terms, the teacher and parents agree that David's handwriting has improved and that it is within an acceptable range for a boy his age and class placement, the intervention plan is terminated and things proceed as they would for all other children in the class. However, if parents and teachers believe that David continues to have handwriting difficulties that are unusual for a boy his age, the teacher will ask the parents if she can contact the Learning Support Teacher. This is Stage Two of the process.

If the parents agree, the Learning Support Teacher will complete an assessment of David's handwriting and other academic skills to try and discover if he needs assistance beyond what the class teacher can provide. The Learning Support teacher will begin to work with David and will complete an educational programme for him, which will also be shared with the class teacher and his parents. This plan will indicate additional strategies that the class teacher can implement and may include strategies for the parents as well. This educational programme will be implemented over one school term and the parents will be kept informed on a regular basis.

If after one school term the plan is successful, the Learning Support teacher will stop helping David and he will continue along with all the other children in the class, without receiving special help. If the educational plan is not successful, the process moves on to the next step (Stage Three) and the parents will be asked if they agree that David should have a more intensive assessment of his difficulty from an educational psychologist or other specialist with expertise in the area of handwriting. If the parents agree then a timetable for the assessment will be put in place and an assessment completed. After the assessment, the parents and teachers will receive a written report from the assessor and a multidisciplinary meeting (see glossary) will be convened to discuss the results. The parents will be invited to this team meeting. The results of this meeting will determine whether or not David requires special education.

If he does, **an Individual Education Plan**, sometimes just called an Educational Plan (see page 12), will be written for David. Everyone involved in David's education, including his parents, will have an opportunity to devise the plan and everyone will be made aware of exactly how David is going to be helped, who will help him, where this help will take place, how often he will receive help and for how long. About one year later the plan will be reviewed.

The purpose of this book is to provide all the essential information that parents and teachers need to enable them to understand this process and all that is involved in implementing it. The reader will find extensive information about the Law, the Department guidelines, the structure of special education, the nature of all the special education conditions that are recognised by the Department, the nature of assessment and what information it yields, and how to use the curriculum to assist all children, especially children with special education needs. The reader can work through the book sequentially or dip in and out of sections as necessary. If the reader is a teacher, or a student in a College of Education or University Higher Diploma in Education programme, the book is an absolute necessity. If the reader is a parent some sections are necessary and some are nice to know. Whatever the case may be this book is intended to be **The Essential Guide to Special Education in Ireland**.

What is special about special education?

What is the greatest speciality the individual teacher brings to his or her work with children having special education needs? The answer may not be particularly scientific – it may even be trite and old-fashioned, but the single most important speciality is acceptance. A teacher who recognises that the child comes with gifts, the gifts of being just whom and what they are, who accepts them as they are, is a genuine 'special' teacher. To be accepted just as you are is the starting point for all educational progress.

This is not always easy but it is the essential starting point for every classroom and special education teacher. Each one must also apply specialist skills and strategies and work patiently with the child at all times. No matter how much post-graduate education a teacher has acquired, if they lack the basic capacity to recognise and accept the gifts of every child, they will do no good, and in some cases may do considerable harm.

PART ONE

Assessment and Educational Planning

The best advice I have for a parent or teacher regarding special education is that it is critically important to be knowledgeable about your legal rights, your entitlements and the seemingly ever-changing structure of special education in Ireland. It is essential to keep current with all developments in this area. Regrettably, those who are not informed will encounter a system they cannot understand and discover that those who are better informed are receiving more appropriate educational services. For parents and teachers of children with special needs, nothing is more important than a thorough understanding of special education in Ireland (the primary motivation for writing this book).

While there are many parents and teachers who are savvy about rights and entitlements (having learned the hard way I suspect!), I fear that most are in a state of near total ignorance regarding the most important *facts* of special education. By the facts I mean the little things that matter the most:

- What is the meaning of a child's special education condition?
- What does an assessment tell me about a child?
- How is assessment information translated into educational practice for a child?
- What sort of educational and therapeutic interventions have been put into place?
  - o By whom?
  - o When?
  - o Where?
  - o How often?
- Is a child making progress?
- How can I tell?
- Is there any evidence to support the assertion that he or she is making progress?

These are the vital questions; the answers inform everything that will be done for the child in the future. The answers tell a story, the story about the effectiveness of a child's special education programme. When it comes to special education, nothing matters more than the child and the programme of intervention that is put into place for the child. And when it comes to the programme, nothing is more important than the ability to accurately assess its effectiveness over a period of time.

Unfortunately, the resolution to issues about appropriate educational provision for children with special needs is often a matter of speculation for parents and teachers alike. Emotions sometimes outweigh logic. Parents, reacting emotionally, believe a programme is not appropriate – the special education staff, reacting logically, believes it is. Without some sort of objective information, progress or the lack of it cannot be adequately calculated. That underscores the requirement in the law that programmes and educational plans be assessed at least annually. However, unless the people involved *understand* assessment, they will not be in a position to make suitable use of the information that the assessment yields.

All concerned with a child's special education programme should be able, at the end of at least one year, to answer the following questions with accuracy:
- How is the child functioning, compared with other children of the same age?
- How is the child functioning, compared with others in the same class?
- How much educational progress has the child made (what has been learned) since the last test battery?
- Has the child made noticeable progress in their special education programme?
- If the child has shown an increase in age equivalent test scores (see page 245), has the child actually fallen further behind the peer group?

Let's begin with the **nature of assessment** itself. How are children referred for assessment and why? What information does an assessment generate and how is that information used to create an individualised educational plan for a particular child. In Appendix One, I provide a lot more detail on psychological and academic assessments.

Assessment

As has been outlined in the Prologue, when a parent or teacher suspect a child has a learning problem in any of the following areas: academic, physical, emotional, social, or behavioural, a short intervention plan must be developed by the classroom teacher. The plan must be implemented over about two full terms before it can be determined to have been successful or not. At that point, a decision is made to either terminate the intervention and close the case (in the case of a successful outcome) or move to Stage Two of the process for additional assessment by a Learning Support teacher. So, within the staged process of special education, assessment begins at the first stage, becomes more specific and diagnostic at the second stage, and involves specialists outside the school at the third stage.

Referral for outside or further assessment is mandatory under this process when the class teacher believes his or her intervention plan has not helped the child. It will be undertaken by the specialist teacher; most often the learning support teacher.

Referral

Children are referred for assessment when a high-incidence (common) special needs condition (specific learning disabilities such as dyslexia) is suspected, or when a low-incidence conditions such as autism, is present. Children with the more low-incidence conditions or with genetic/ chromosomal conditions are frequently assessed before they get to primary school, because the condition is so obvious.

Referral for more specific assessment (to be performed by specialists outside the school, such as psychologists or speech and language therapists) is usually processed within the school, unless the child is not of school-going age. It goes through the special education support team, to the principal, on to the SENO and must only proceed with parental permission. No child can be assessed beyond the screening measures undertaken by a classroom teacher without parental permission. When parents refuse permission, the school can make a petition to the Circuit Court to get an order that requires the assessment.

The most commonly sought-after assessment is a **psychological assessment**. A psychological assessment is nearly an absolute requirement for making a determination about the presence or absence of a special education condition. Psychologists have been assessing human intelligence since the turn of the century and are the only professional group who can do so. The use of intelligence tests to classify children into various categories of special education need has been an area of controversy for years. The reasons for this are complex and relate to some erroneous thinking about intelligence, which will be explained further on.

In Ireland, psychologists from the National Educational Psychological Service (NEPS, see page 190) complete most educational assessments. After speaking with a teacher, parents, a special education school team and perhaps the SENO, the psychologist will work individually with the child and administer a test. The results of this test will generally classify the child into a category of intelligence such as average, below average, gifted, learning disabled, etc. In Ireland the result of the psychological assessment has an enormous influence on the type of special education service a child will receive and where they will receive it.

Psychologists will sometimes administer **academic tests** to children to determine their facility in reading, spelling, mathematics or written language. There is a perception that the results of a psychological assessment are sacrosanct, results so enormously true that they can't be questioned. In what follows I will outline what I believe, as a psychologist, constitutes a reasonably valid psychological assessment process. Assessments with psychological tests and measures are an important part of the special education process but the results must be interpreted intelligently.

Comprehensive referral information

The psychologist who is assessing a child should have a clear and unambiguous understanding of why they are undertaking the assessment. This means they need accurate referral information. A referral that indicates a child is having 'reading difficulty' or 'behavioural problems' gives the psychologist little useful information to work with. What is needed are the specific facts: the aspect of reading that the child is having difficulty with (sounding out words, understanding what he has read, remembering what has been read and so on). If the referral relates to behavioural issues, then specific information is needed about where the problem arises (what time of day, with whom, under what conditions, during which subject, how often etc.). A psychologist needs to make every effort possible to intelligently interpret the referral information and reflect carefully on how to proceed.

Interviews

Following this careful analysis of the referral information, the psychologist must talk with parents, teachers and others involved in the child's life. Interviews with parents are critically important because they yield significant information about the child's birth and developmental history, health history, family background and incidence of learning problems in other family members. Additionally, if the concern is behavioural, the parents will be able to inform the psychologist if the problems arise in the same set of circumstances as seen in school.

The parent interview will also be a time for the parents to ask questions of the psychologist, such as 'What information do you expect to discover in your assessment?', 'What use will be made of the information you discover?', 'How will it relate to the educational plan that is drawn up if one is needed?', and so on. A careful parent interview, and interview with other family members, is the best starting point for any assessment.

The psychologist will also need to speak with the classroom teacher and any specialist teacher(s) involved in the child's education. It will be necessary to discover the exact nature of the teacher's concerns, if this hasn't already been done in the referral phase. The psychologist will need to know what interventions the teacher has tried, what changes in methodology have been implemented, what different learning resources such as books or videos have been used, what sort of different groups have been tried with the child and peers and so on.

Importantly the psychologist will need to inquire about the child's strengths and talents. What is the child best at doing and learning? Under what conditions does the child show a keen interest in learning? Does the child learn differently while there is whole-class teaching as opposed to when there are active group-learning activities? Is there a difference in learning style when the child is tackling different content areas (reading time versus geography)? The psychologist will also want to inquire about attendance and punctuality, or any interruptions to schooling and their cause (frequent family relocations, illnesses etc.). The answers to these and many other questions may shed significant light on the child's alleged problems.

Observations

After the interview stage comes the observation stage. The psychologist should observe the child in multiple settings, depending on the nature of the referral concerned. If the concern relates to a single subject area such as reading, the psychologist will obviously need to observe during reading time. But children read at other times during the school day, and the psychologist must take every opportunity to observe the child in as many different contexts as possible.

This is especially true when the referral issue concerns behavioural problems. Then the psychologist must observe the child during structured in-class activities, unstructured in-class activities, physical education, yard time, free playtime, in the family home and out in the community with family members etc. The more opportunities to observe and carefully record the child's behaviour under multiple conditions, the more intelligent will be the results of the assessment.

After completing the interviews and the observations it may be necessary to administer tests to the child. I say it *may* be necessary for a variety of reasons. If the information gathered indicates that the alleged learning or behaviour problems are a result of frequent school absences, poor teaching, a significant mismatch between a child's learning style and a teacher's teaching style, it would be erroneous to assume the problem lies within the child. Under these circumstances the outside variables need to be remedied, not the child. All psychological assessments should proceed under what is sometimes called the 'null hypothesis', that is, there is nothing wrong with the child until incontrovertible evidence has been accumulated to indicate there is something wrong. It may well be a fact that there are as many teaching or schooling disabilities as there are learning disabilities and it is the sometimes-unfortunate role of the psychologist to get to the truth.

The Legalities of Assessment: Who and How?

An assessment may be carried out by professionals deemed to have necessary expertise and qualifications by the Health Executive or the NCSE (see page 188), and might include:
- A psychologist
- A medical practitioner
- The principal of the school which the child is attending, or a teacher of that school nominated by the principal
- An appropriately qualified social worker
- A therapist who is suitably qualified to provide support services in respect of the special educational needs of the child

Parents are to be involved at every step in the process and it is up to the NCSE, the Health Executive, or the principal to facilitate their involvement. Parents are to be given notice of intent to carry out an assessment, and no assessment can be undertaken without parental permission, in writing.

If a parent refuses to give consent, the Health Executive or NCSE can apply to the Circuit Court for an order that dispenses with the requirement for parental consent. The court may make such an order if it feels it is within the child's best interest to do so.

When undertaking an assessment, the Health Executive or NCSE shall take into consideration any relevant previous assessment of the child that is available to them at that time. After an assessment is completed, a statement about its result is written and provided to the parents. The results will also be provided to any such persons engaged in the education of the child as considered appropriate by the Health Executive or NCSE. The parents will be notified of this fact and be provided with the identity of the persons to whom it has been given.

The Education for Persons with Special Education Needs Act (see page 153) gives parents the right to appeal the results of an assessment to the Appeals Board of the NCSE, under the grounds that it was not carried out under the terms outlined in the Act. The board will make a determination of the appeal with two months.

We have an assessment, now what?

The results of an assessment, combined with all the observational and other information that has been learned about the child are used to make some important decisions:

- Does the child have a special education condition?
- If the child does not have a special education condition, what supports will the classroom teacher and learning support teacher (if one is needed) institute to help the child?
- If the child does have a special education condition what is that condition?
- What sort of special education is required to remediate that condition?
- Where will that special education take place?
- How will the information generated by the assessment be used to create a focused educational plan that is individualised to the unique learning profile of the child?
- What information will be gathered to assess the impact of the educational programme after one year?

The validity and reliability of assessment

All assessments are designed to measure something. It might be intelligence, reading achievement, social skills, or hundreds of other pieces of information that will be used to help design an appropriate educational programme. The more skilful you are in measuring, the more useful the information you will gain from the measurement. Therefore it is important to know for sure that a particular form of assessment is valid – that it actually measures what it is supposed to measure.

All tests and assessments distributed by publishers contain information about how valid the test is. It is the responsibility of the assessor to know for sure whether or not the assessment method being used is valid for the purpose of the assessment. For example, if we weigh a watermelon and use its weight to estimate how sweet it will be, we are using an invalid assessment technique because weight is not related to the sweetness of a melon. Likewise, a test of intelligence is not a valid measure of whether or not a child has an autistic spectrum disorder.

Reliability

Tests measure things, specific things, and sometimes tests are used to measure the same thing several times. If we are going to use a test to measure something, we need to know if the test will produce similar results each time it is used. In other words, how consistent is this test in measuring what it is supposed to measure. A test that yields highly variable results is not reliable and therefore cannot be considered an appropriate assessment tool. Imagine if we put a 25-kilo weight on a scale every day, and got different readings each day. How could we understand the results and make decisions based on them?

A final word ... a test can be reliable but not valid. Suppose for example that a teacher tells her students that the weight of their hands will predict who will run the fastest from one end of the yard to another, and that the child with the heaviest hands will be the fastest runner. Each day for one week the teacher weighs each child's hand. From day to day the weight of any child's hand will not change significantly at all. Therefore the test being used (hand weight) is reliable because it gives the same results each time it is used. However, it is not valid because speed in running is not determined by the weight of your hand and it could well be that the child with the lightest hands is the fastest runner. Therefore it is an inappropriate measure yielding useless information. Make sure to ask questions about how your child was assessed and don't be concerned about asking questions related to the reliability and validity of the tests that were used.

Assessment and Educational Planning

There is nothing more important than the **Individual Educational Plan (IEP)**, or Educational Plan as it is called in the Education for Persons with Special Education Needs Act, 2004 (I will refer to it by its proper name, the IEP). The IEP is the heart and soul of the education of a child with special education needs. Without an appropriate, focused IEP it is impossible for anyone involved in the education of a child with a special need to know if the interventions put in place are working.
There may be education in a so-called 'special' setting but it is not, and cannot be, special education.
So important is the IEP that it can be stated with certainty that where there is no IEP for a child receiving special education, there is no special education. For an excellent publication concerning IEP's, which contains information that is relevant to parents and teachers, please visit the website of the National Disability Authority at: www.nda.ie

Parents should always be involved in the writing of the IEP. It may not be necessary for them to sit for several hours while a specialist teacher, the classroom teachers and a psychologist begin to draw up the plan, but it is critically important that they be involved, and fully involved in the final drafting of the plan. It is also imperative that they understand the plan, its aims, goals and objectives. They also need to know how it will be evaluated, when it will be evaluated and how it can be changed if it needs to be changed. At secondary school, it is also important that the student himself or herself be involved in the development of the IEP. Special education at secondary level works best when the most important person in the special education equation is directly involved in the development of the programme.

What should be included in an IEP?

The Department of Education and Science, along with the National Council for Special Education have not drawn up a standard template for what an IEP looks like at the time of writing. Whatever the look and feel of the final IEP document, it should, as a matter of best international practice, contain all the following elements in some form or other:

1. A statement, based on individual assessment, of current level of functioning in each of the following areas:
   - Academic
   - Social
   - Behavioural
   - Perceptual
   - Physical
   - Communication
   - Life skills

2. A list of annual goals that can reasonably be expected to be accomplished by the child over an academic year. There must be a direct correlation between the annual goals and the present level of educational functioning. In addition:
   i. All goals must relate to measurable, observable behaviours. They must address the manner or level that the goal will be met and under what circumstances the goal will be met.
   ii. Goals must be realistic in terms of present level of functioning and overall cognitive abilities.
   iii. Goals must be prioritised on the basis of age.

3. A statement of short-term objectives, in a sequence of steps that move towards annual goals.

4. A statement about the evaluation criteria, that is, the level of performance necessary for a positive assessment of 'mastery' of each of the objectives.

5. A statement of the frequency of assessments, that is, the schedule of evaluations to determine progress towards annual goals.

6. A statement of the evaluation procedure that specifies the manner in which each skill will be assessed.
7. The IEP will state who is responsible for the services required.
8. The IEP will state where each required service will take place.
9. The IEP will state the frequency and length of each educational intervention that is, how often the intervention will take place (once a week, twice a week) and for how long will each intervention last (one-half hour, one hour).

Parents must be involved, not just informed

All IEP's are pre and post-tested annually. Parents must be involved in true collaboration in the writing of an IEP. 'True' collaboration does not consist of being called into school, handed an IEP and being asked to look it over and agree to it! This is being 'informed'; the development of an IEP requires that parents be 'involved', not just informed. Any IEP that is not shared with parents and is not written in collaboration with parents is inappropriate special education provision.

As stated, the IEP should contain statements of annual goals plus short-term objectives related to each goal. This should be done for every area in which the child is unable to benefit from regular education without special assistance. Typically, it is recommended that there should be four short-term objectives for each annual goal (and there should be between four to ten annual goals). Too many goals and short- term objectives make a plan cluttered. A small number of key goals and objectives makes it possible for the special education team to assess the effectiveness of interventions and monitor progress.

When planning interventions, the team must take into account the child's present skill level, the teacher's skills, the resources available and the likelihood that the intervention will actually be implemented. There is no point in listing goals, objectives and interventions that cannot reasonably be implemented. For example, there is no point stating that a child will receive individual speech and language therapy twice every month for one school year, when no speech therapist is available.

Seven steps to a successful IEP

Hammer and Malatchi* have adapted the popular book (by Stephen Covey) about the habits of successful people and related it to seven successful ways of writing an appropriate IEP:

1. Be proactive – all should adapt a collaborative, facilitative and partnership attitude.
2. Begin with the end in mind – learn about the child, predict reasonable outcomes.
3. Put first things first – prioritise, you can't work on everything in one year.
4. Think win-win – all should benefit from and feel good about the IEP.
5. Seek first to understand, then to be understood – listen to parents, then share your point of view.
6. Synergise – have a multi-disciplinary team that works together.
7. Sharpening the saw – seek continuous improvement in the plan from one year to the next.

Children's behaviour

A few words about children's behaviour as it relates to IEP planning are necessary at this point. Children with behavioural problems, who are receiving special education services, do not always present with academic deficits. If they do, then the academic deficits must be remediated through the IEP process; if they do not, then the behaviour will be the sole focus of the IEP. There are a number of helpful questions the IEP team can consider when devising a plan for a child with behaviour difficulties:

1. Does the child need to learn new behaviours they don't already have?
2. Does the child need to have behaviours eliminated?
3. Does the child present with unsafe behaviours or behaviours that disrupt the learning environment of others?
4. Does the child's current behaviour profile require a behaviour intervention plan?
5. Is the child frequently removed from the mainstream classroom because of behavioural difficulties?

---

* Hammer, E. & Malatchi, A. Seven habits of highly effective IEP teams.

6. What has been the impact of this removal on the child's academic skills?
7. Is the child's behaviour directly related to a special education condition?
8. What interventions have been implemented in the mainstream classroom and what was their effect?
9. How do parents or other family members deal with these behaviour patterns, if they occur at home, and how successful are those interventions?
10. For how long has the child exhibited these behaviours?
11. Where do the behaviours occur? What time of day? What day? With whom?
12. Are there times and places where the behaviours seem worse?
13. Are there times and places where the behaviours do not occur at all?
14. Is the child remorseful about the behaviours?
15. Is the child aware that he or she is making a mistake in behaviour?

The key question in writing educational objectives is: 'If the service we are providing is effective, what changes in learning or behaviour in the child will we actually **see**?' The answer to this question helps us write objectives that can be easily assessed. For example, if a child with significant dyslexia has an IEP that states 'Colm will improve in reading this year', it is impossible to measure. What is 'improvement' in this context? How long is a year, is it an entire calendar year or a school year? A better statement of this objective might be 'Colm will be able to recognise a minimum of forty-five words out of sixty when presented with flash cards at the end of this calendar year'.

Similarly, for a child with autism who's IEP has an objective that states, 'Judy will play some simple games with other children', it will be hard to accurately measure. This objective could be written, 'Judy will share the trucks and cars with one other child when engaged in a cooperative sorting and counting activity by the end of the calendar year.'

Because the IEP is so important, it is imperative for special education teachers to be fully knowledgeable about its creation and implementation. Additionally, IEP's are dynamic documents which need to be reviewed at least once a year but might be reviewed more often if necessary. It is not always possible to create goals and objectives that can be met and for this reason it is the responsibility of parents and teachers to keep a watchful eye on the child's progress in relation to the IEP and re-convene the team if necessary. There is no excuse for waiting until the end of a school year when there is adequate reason to believe things are not going well.

The reader is advised to review the section of this book about the role of the learning support teacher (see page 195) in order to become familiar with the nature of their particular methods of educational planning.

PART TWO

Special Education Conditions

The Department of Education and Science recognises the following special education conditions as being 'entitled to all the necessary supports and resources required to provide a child with an appropriate education':

- Physical Disability
- Hearing Impairment
- Visual Impairment
- Emotional Disturbance and/or Behavioural Problems
- Borderline General Learning Disability
- Mild General Learning Disability
- Moderate General Learning Disability
- Severe/Profound General Learning Disability
- Autism/Autistic Spectrum Disorders
- Specific Learning Disability
- Specific Speech and Language Disorder
- Children with Special Education Needs Arising from an Assessed Syndrome
- Multiple Disabilities
- Exceptional Ability/Giftedness

Within these general categories, some additional conditions may carry the same entitlement to resources and supports, such as Dyslexia and Attention Deficit Hyperactivity Disorder. The DES has specific guidelines for the assessment of these conditions that will be outlined below.

The reader is cautioned that the names of special education conditions differ from one country to another. For example, the Irish category of General Learning Disability is called "Mental Retardation" in America. In America what is often called "Learning Disability" is called "Specific Learning Disability" in Ireland. The confusion in the names of conditions can result in parents or teachers accessing incorrect information about educational strategies, particularly on the Internet. Make sure you know what you are reading about and how it matches the Irish categories.

Let's look at each category in more detail. Note that in the fourth section of this book I have included some general information about the conditions and I describe some interventions for each category of special education need.

Physical Disabilities

This category of special education condition refers to children with conditions such as cerebral palsy, Spina Bifida, muscular dystrophy, or children who suffer accidental injury which limits locomotors or motor function and who require special intervention and support to avail of education. These supports can range from the use of wheelchairs or special seating, to the use of adaptive computer technology. This category includes a condition called 'oral dyspraxia' which interferes with a child's ability to communicate verbally.

Children with physical disabilities can often be educated in the mainstream school. However, if the disability is so severe that the child requires physical care such as feeding, toilet-assistance, physiotherapy, occupational therapy or nursing assistance on a regular daily basis, the child may be referred to a special school which has all the necessary staff and equipment to support the child's educational programme. A good many children with physical disabilities have been assigned a Special Needs Assistant (SNA), whose role it is to provide feeding and toileting care, who see to it that they can negotiate the physical environment of the school and can integrate into the whole life of the school. Remember that a SNA works under the direct supervision of the classroom teacher and is not to assume direct teaching responsibilities.

A common misunderstanding is that children with physical disabilities also have intellectual disabilities (that their intelligence is well below average). Although it is true that some conditions of physical disability may have an impact on intelligence, it is not necessarily the case. For example, children with cerebral palsy may often be unable to speak intelligibly or to speak at all without assistive computer technology or other equipment. However their intelligence can be well above average. One should never make the mistake of assuming anything about a child or adult with a physical disability. It is the critically important role of assessment to make informed judgements about the intelligence of these children.

Another misunderstanding about children with physical disabilities is that their mobility difficulties inevitably restrict and limit their participation in the life of the school. This is of course untrue and all children can participate, to the best of their ability, in every aspect of education, including physical education, class trips, assemblies and the day-to-day routine of the classroom. The advances in technology now make it possible for every child and adult to communicate their needs, desires, hopes and interests and it is the right of these children and adults to have access to whatever is necessary to allow them to do so.

Hearing Impairment

Children with hearing difficulties serious enough to impair their capacity to hear and understand human speech, or which prevents them from participating fully in classroom interaction and interferes with their learning, will require support in mainstream classes. This category is not intended to include children with mild hearing loss that can be adequately corrected with hearing aides. Children with hearing impairment who are included in mainstream classrooms benefit from the specialist support services of the 'visiting teacher' for the hearing impaired.

There are levels of hearing impairments that are so severe that they limit a child's ability to receive an education in a mainstream school. For these children, there are special schools for the hearing impaired, with staff that are trained in sign language and other adaptive methods to instruct and help such children. There is disagreement in the parent community about what constitutes an appropriate education for children with severe hearing loss, given the international movement towards 'inclusive' education. Many parents are unwilling to have their children educated in segregated settings in special schools. As is the case with physical disability, improvements and advances in technology are making it increasingly possible for children with severe hearing loss to communicate effectively.

Visual Impairment

This category covers children who have a visual disability that is so serious it impairs their capacity to see; therefore they will have significant difficulty in mainstream classes. This category is not intended to include children whose vision is corrected satisfactorily by wearing spectacles and/or contact lenses. Just as there are special schools for children with hearing impairment, there are special schools for children with visual impairment. However, it is the case that more and more children with visual impairment are being educated in mainstream classrooms. These children also benefit from receiving the specialist services of the visiting teacher for the visually impaired.

Emotional Disturbance and/or Behavioural Problems

This category of special education condition refers to children who are being treated by a psychiatrist or psychologist for conditions such as hyperactivity, Attention Deficit Hyperactivity Disorder, conduct disorder or other problems that disturb behaviour. It does not include children whose behavioural difficulties are adequately dealt with by ordinary school procedures on discipline. This category, sometimes referred to as simply EBD (Emotional and behavioural disturbance), is one of the most misunderstood categories of special conditions and will require lengthy discussion.

The category of 'emotional and behavioural disturbance' is probably the only special education condition for which the person who has it is actually blamed for having it. Frequently referred to as 'cheeky', 'bold', 'messers', or 'delinquents', children and adolescents with behavioural or emotional problems are a high profile concern to everyone in the system, especially their parents. Their behaviour can disrupt a class or even a whole school and can provoke adults into wanting to punish them. In severe cases, it can provoke the system into removing them as soon as possible, especially at secondary level. Currently being referred to as presenting 'challenging behaviour', these children can be a puzzle to all concerned. (There is a further article on challenging behaviour on page 69). A basic understanding of their behavioural pattern can go a long way to improving the entire situation. Educational programming for these children must be well designed and often requires specialist intervention from outside of the classroom.

Let's examine the area of emotional and behavioural disturbance in general. This will be followed by a more detailed discussion of attention deficit hyperactivity disorder.

What we call emotional and behavioural disturbance can result from a variety of causes, some of which are neurobiological (disruption of the electrical-chemical activity of the brain) and some of which might be environmental (the result of the child's circumstances/home environment). Among the neurobiological causes of emotional and behavioural disturbance are the serious psychiatric illnesses such as clinical depression, anxiety, phobias, childhood schizophrenia, bi-polar disorder (manic depression) and other severe conditions. Although affecting few children in number, these conditions all require psychiatric treatment and all will interfere in a significant degree with learning, adjustment and behaviour in school. Fortunately most of these severe conditions can be successfully treated with a combination of medication and counselling.

The most important thing for teachers and parents to consider is that these illnesses are just as much medical in origin as is diabetes or the flu. The causal agent may not be as easily recognisable as a malfunctioning pancreas or a virus, but the end result is the same – a system out of balance with itself and in need of medical intervention to restore that balance.

What are more puzzling are the more severe and disruptive conditions of children's behaviour that affect children and adolescents on a frequent enough basis to merit mention: Oppositional Defiant Disorder and Conduct Disorder, the two most severe and disruptive conditions of children's behaviour.

Oppositional Defiant Disorder (ODD)

This condition often appears early in life and can be distinguished from the usual tantrums of the young child because *it doesn't seem to go away.*

A child with ODD will refuse to comply with even seemingly reasonable parental and teacher requests, will do the exact opposite of what was asked, or stubbornly refuse to comply at all. This pattern of behaviour can only be fixed over a long number of years and is usually not affected by the ordinary disciplinary methods of parents and teachers. What works on other children does not seem to have any impact on a child with ODD; instead they escalate the behaviour, the child becoming *more* stubborn and resistant to change. The child will become more provocative and defiant, will have angry outbursts and use harsh words and will seem oblivious to every sort of punishment meted out to them. At home they refuse to clean off the table, to go to bed, to do homework, to help care for a brother or sister, or to make a trip to the shop to get some milk. In school they refuse to listen to teacher when told to take out a copy, read a book, stand in line or remain quiet while instructions are given.

This pattern of behaviour seems willful and a direct challenge to any source of adult authority. As these children get older, the severity of their opposition can get worse and they can be prone to lash out when confronted with direct orders or commands, even to the point of being physically confrontational to those they love most. There is no known cause of ODD as yet, although contemporary theories seem to revolve around 'functional brain problems'. One thing is certain … bad parenting does not cause ODD. Inconsistent or misdirected parenting may cause children to be defiant and stubborn some of the time but these children usually respond to corrective measures. The child with ODD rarely responds at all.

Conduct Disorder (CD)

This is the single most disruptive and difficult condition to affect the behaviour of children and adolescents. Usually not diagnosed until the child is an older adolescent, often age eighteen, conduct disorder seems to be related to what at first appeared as ODD. Conduct disorder is different from ODD because it is more severe and has more socially disruptive and disturbing characteristics. Like children with ODD, children with CD are stubborn to the point of absolute refusal – they are provocative and challenging to any adult in a position of authority over them. But unlike children with ODD, they can be physically aggressive, threatening and menacing. They are likely to destroy property, break the law in petty criminal ways (sometimes involving more serious criminal behaviour), extort money from others, lie, bully, cheat and steal. Children with CD often have little or no ability to empathise with others, particularly those they have wronged. The child or adolescent with CD is much more difficult to manage than the child with ODD and will most often require education in a specialised setting with strict behavioural limits, including restrictions on movement and freedom until privileges are earned.

The long-term outcome, even with treatment, for children with CD is not particularly encouraging. They do not respond well to counselling or psychotherapy, which is why behavioural management and treatment is both required and recommended. Many of them cannot be educated in a mainstream setting because of the severity of disruption they cause to those around them in the classroom/school. Like children with ODD, the cause of CD is not known at this time. Although environmental factors play an important role, there is increasing evidence that disruption of the emotional and empathetic centres of the brain are involved in the disorder.

Attention Deficit Hyperactivity Disorder (ADHD)

Enough is now known about Attention Deficit Hyperactivity Disorder (ADHD) to be certain that it is a neurobiological disorder that seems to affect about five to seven percent of all children. Descriptions of children with ADHD have been around for over one hundred years in the medical literature, although the current terminology is new. Previously these children might have been referred to as having 'defects of will', as having 'minimal brain damage' or 'nervous conditions'. Whatever the name, ADHD is a well-recognised condition and although some authors contend it is nothing more than an alternative way of behaving, anyone who is a parent or teacher of a child with ADHD knows that these children are different from other children.

ADHD can be distinguished from normal levels of distraction, behaviour or inattention by:
- Having first been present before the age of seven
- Having lasted for more than six months
- Causing severe disruption in at least two of the following three areas of the child's life: school, home or community

There seems to be a genetic component to ADHD and it is not unusual to discover a family pattern to the condition. Children with ADHD are noteworthy for their inattention on tasks that require sustained attention. Interestingly, they can be quite attentive to other tasks, such as video games or playing with Lego. They will have poor impulse control (act without thinking) which has caused ADHD to be sometimes referred to as the 'ready, **fire**, aim' syndrome. They are poorly focused, easily distractible and often highly overactive. More boys than girls seem to be diagnosed with ADHD, except in the category of 'predominately inattentive' (see below), where the gender breakdown is more fifty-fifty. The number of boys with *severe* ADHD exceeds the number of girls to a significant extent. Although parents have been historically blamed for 'having' children with ADHD, it is clear from the research that ineffective or bad parenting *does not cause* ADHD. There is no racial or ethnic difference in the number of children with ADHD and it seems to affect children at nearly every level of intelligence.

There are three subtypes of ADHD, each with its own distinctive profile of behaviour and intellectual functioning:
- Predominately Inattentive Type
- Predominately Hyperactivity/Impulsive Type
- Combined Type

Children with **Predominately Inattentive ADHD** are day-dreamy, poorly focused, appear to have memory problems, don't seem to listen to instructions and often have difficulty completing even simple tasks because they somehow get 'lost along the way'. Children with **Predominately Hyperactive/Impulsive ADHD** are prone to act without thinking, do whatever comes into their mind and will answer questions before the adult has finished asking them. They have difficulty abiding by the rules of games, seem driven from one activity to another without completing any activity and in general, appear to be highly fidgety and restless, as though they are squirming out of their skin. These children cause the greatest disruption to those around them and are frequently in difficulty with their peers for activities like pushing, shoving, kicking or fighting. Unlike children with conduct disorders however, the child with hyperactive/impulsive ADHD *does* feel remorse and regret for what they have done wrong, genuinely wants to do better in the future and tries to improve their behaviour only to fail again, sometimes over and over.

Children with the **Combined Type of ADHD** have all the symptoms of inattention, hyperactivity and impulsivity. This subtype of ADHD causes the greatest interference with school and home adjustment. From the descriptions above, it is easy to see that children with ADHD are frequently at odds with the adults and children around them. For these reasons they are often rejected as play partners, not invited to other children's homes, frequently punished at home and at school and just do not fit comfortably into the world around them. It can be a lonely place to live, especially when you want to change your behaviour but seem doomed to make the same mistakes over and over. It is no wonder then that ADHD, left unrecognised and untreated, can cause life-long difficulties such as depression, poor self-esteem, lack of self-confidence and learning difficulties.

Some thoughts about ADHD

A few things need to be said about the nature of ADHD to avoid the common confusion surrounding the condition. Children with ADHD *did not choose* to have this disorder and it is not their fault they have it. They genuinely do want to behave better and learn better but their brain is working in such a way as to prevent this desire from becoming a reality. ADHD does not *cause* antisocial behaviour, although antisocial behaviour can be associated with ADHD. There is no direct link between the two and it is most often the case that any antisocial behaviour is more a result of unrecognised and untreated ADHD and its secondary effects such as social problems, depression, poor impulse control, and low frustration tolerance. Children with ADHD are not less intelligent than other children and children at almost all levels of intelligence can have ADHD (there is some disagreement about whether or not ADHD can be present in children with severe autism or significant general learning disabilities).

ADHD can be successfully treated and the primary places of treatment are home and school. The earlier that ADHD is recognised and the earlier it is treated, the better the outcome will be.

ADHD is commonly associated with other special education conditions such as Specific Learning Disabilities (dyslexia), ODD, CD, or a movement disorder called Tourette's Syndrome. When two or more conditions exist simultaneously, they are said to be 'co-morbid'. Co-morbidity is common in ADHD and when it is, all the conditions that are present must be treated and addressed at the same time, in order to achieve a more successful outcome. This makes treatment at times complicated and may require the services of a number of professionals (a psychologist and psychiatrist working in consultation with a special education teacher perhaps).

The importance of an understanding and sympathetic classroom teacher cannot be overstated. It is likely that in every classroom in Ireland there is at least one child with ADHD. An understanding of the condition is an absolute necessity on the part of every teacher at both primary and secondary level. ADHD is a lifelong condition that will require periods of treatment and remediation throughout the life span. It does not go away, but it need not be a debilitating condition.

The issue of medication for children and adolescents with ADHD is a touchy one for many parents. ADHD has been treated with medication, the most common one in use being Ritalin, for well over thirty years now. In the United States there is considerable controversy about the large numbers of young children treated medically for ADHD, but this is not an issue in Ireland at the present time. Children with severe ADHD, particularly the combined type or predominately hyperactive/impulsive type, can be a danger to themselves as a result of impaired impulse control. A parent may have to choose between having a child who can put his life at risk on the one hand and providing medication to the child to reduce this risk.

Medication for ADHD does seem to have a positive effect on learning as well but this is more a result of better attention and concentration than anything else. In any event, it is a family decision and one that must be discussed carefully and thoughtfully with a child or adolescent psychiatrist. For all but the most severe forms of ADHD, every known system of behaviour management must be exhausted before considering medication. Although a last-resort solution, medication does have proven benefits for children with severe ADHD.

The General Learning Disabilities

General Learning Disabilities (GLD) are a group of conditions which all have one thing in common; the child will have a level of intelligence, (as measured by an individually administered intelligence test by a psychologist), that falls below the average range (90-110 standard scores; 25th-75th percentile range). It is this low level of intellectual functioning that results in interference with the rate and process of learning in the classroom and which makes these children eligible for special education services.

The GLD's are presented below in an order of limitation, that is, those with less limited intelligence will be outlined before those with more limited intelligence. The reader is cautioned to re-read the section about intelligence and compare it with the information outlined below, before rushing to any conclusion about its impact on functioning and behaviour.

Another important **note of caution** is necessary at this point, both for parents and teachers. In Ireland, the term General Learning Disability describes a group of children that in the United States are referred to as "Mentally Retarded". This distinction in terminology is critically important because so many parents and teachers use the Internet to get information about Special Needs Education. For example, when accessing a website from America that talks about 'Learning Disabilities', it is referring to children who in Ireland are classified as having a 'Specific Learning Disability'. Interventions that are recommended for these children are in fact inappropriate for children with general learning disabilities as we classify them. The reader is urged to use the utmost caution and to be aware of the country of origin of the website before exploring or implementing strategies that could cause harm or do no good.

Borderline General Learning Disability

This interesting group of children will have a measured level of intelligence (IQ) in the range of 70-79 (2nd-8th percentile). If they are experiencing persistent failure in the classroom, they may be eligible for special education services, most often from the learning support teacher (who works in cooperation and consultation with the classroom teacher, using similar methodologies, strategies and resources). The reason that special education should be provided for them is that we know that many of these children learn quite well when the methods of instruction are appropriate for them; when the teacher is patient and takes things one step at a time and when they are provided tasks which match their ability level.

In order to access more resources for these children, the school must supply evidence to the Department that the staged approach to special education has been implemented along with evidence of an appropriate Individual Education Plan (see page 12).

Children with borderline GLD don't look different from other children as a rule and when not in school, they do not show the type of difficulties that children with more severe GLD's display (known as adaptive behaviour deficits). They act and behave quite like all other children in fact. What becomes apparent in school though, is their difficulty keeping up with the learning pace of other children, their need to have information presented to them several times before they master it, and their need for more concrete and active learning activities. It is not the case that they cannot learn – it's just that the pace at which they learn is slower than that of their peers.

The focus on mixed-ability teaching in the revised primary school curriculum is a benefit for these children because research has clearly indicated that children with borderline GLD learn best in the company of peers whose ability can be the same or more advanced than theirs. Children with borderline GLD do not usually require special schooling and the sort of learning support intervention they require is often best supplied by working in the regular classroom with a small group of children, as well as specific work, in a small group format, in the learning support teacher's classroom from time to time.

Children with borderline GLD can develop behaviour difficulties, because if their condition is not recognised, they are put under pressure to learn at the same pace as all other children. These children were in the past referred to as 'slow learners', a term that is now considered derogatory but which is perhaps the best description of their learning style. Although their progress can be slow and laboured, given support and appropriate teaching they will master most of the primary school curriculum. They may not be able to think as abstractly as other children and they may not be able to draw generalisations or remember and recall as much factual information as other children. It is important for parents and teachers to recognise that some children who are making slow progress in primary school may have a borderline GLD and knowing when, and who, to refer for assessment, can make all the difference in levels of future success.

Mild, Moderate and Severe General Learning Disability

Children with mild GLD (and children with other more significantly severe GLD's) have what are called 'deficits in adaptive behaviour'. Adaptive behaviours are those basic skills we use everyday that we didn't go to school to learn: tying our shoes, putting them on the right feet, knowing how to boil water or cook a simple meal, going to the right shop to buy milk or a hammer, washing our clothes and our bodies; in other words the skills necessary for living our lives successfully at the most minimal level. When the level of both intelligence and adaptive skills are impaired, there is a need for more concerted special education.

Children with mild, moderate, and severe/profound GLD's require us to redefine our understanding of what education is all about. Taken in the narrow sense, education is about learning to read, write, spell, calculate; gaining information about the world, its history and its peoples. Although all of these are important things to learn, children with GLD's beyond borderline will require specific education to help them to live in the world and be able to care for themselves to the best of their ability, in so far as is possible. Therefore their education will plan for necessary skills such as dressing, zipping, fastening buttons, learning to make the tea – whatever skills are deemed the most necessary and appropriate, based on their level of adaptive skills.

The amount of adaptive skills built into an educational plan will be entirely dependent on the profile of strengths and weaknesses that the child presents. The education of children with GLD's below the borderline range will focus on 'living skills' as much as it will on factual skills.

Mild GLD's

Children with Mild GLD have been individually assessed by a psychologist and found to have a level of intelligence (IQ) that is significantly below average, within the range of 50-68 standard scores, below 1st percentile (see page 245). Sometimes these children have syndromes (such as Down Syndrome) that result in this low level of intellectual functioning. Sometimes it can be a result of brain damage caused at birth, or other trauma or illnesses, but often there is no apparent cause.

Many children with mild GLD's can be successfully educated in mainstream classrooms with the help of resource teachers. It is not at all uncommon these days to see children with Down syndrome in mainstream classrooms, learning at a pace that is comfortable for them, and participating in the entire life of the school with success. In Ireland there are still a large number of special classes and schools for children with mild or more severe GLD's. These schools are usually the sites of considerable expertise in teaching these children, although more and more parents do not wish to choose this option of placement. I think it's a great idea for mainstream teachers to make contact with the staff of these schools as they can be an invaluable resource – most will respond favourably when asked for advice on planning and teaching in the mainstream context.

Moderate GLD's

Children with moderate GLD's have been individually assessed by a psychologist and found to have a level of intelligence (IQ) that is within the range of 35-49 standard scores (well below the 1st percentile). These children will present more severe adaptive behaviour deficits and more severe learning needs across all areas of the curriculum. In Ireland, most of these children are placed in special classes or in special schools, although this is a profile that is likely to change over time as more inclusive practice comes into place.

Children with moderate GLD's will frequently display behavioural difficulties as well as their intellectual deficits. The education of these children requires special skills and techniques unique to their learning profile. They should not be placed in a setting for children with emotional and behavioural disturbances just because their behaviour is challenging. Children with Moderate GLD's can learn when provided with an appropriate curriculum, appropriate teaching strategies, and a concerted and cooperative effort among parents, teachers, and others involved in helping them grow and develop to their maximum potential.

Severe/profound GLD's

Those children with severe/profound GLD's have a measured level of intelligence (IQ) that is in a range below 35 (significantly below the 1st percentile). They display significant adaptive deficits and it is not uncommon for them to have physical disabilities as well, frequently with severe communication problems too. Their education must focus on enabling them to communicate their need to be toileted, fed, bathed, comforted, etc. Educational plans for these children are written with small, consistent steps of improvement which will one day allow them to live lives which may not be unsupervised and independent, but that will permit them the necessary personal dignity and freedom we all enjoy: interaction with peers, access to recreation, training for appropriate work and the need to manifest love and receive love in return.

Autism/Autistic Spectrum Disorders

These children have been assessed by a psychologist or psychiatrist and classified by recognised criteria as having one of the Autistic Spectrum Disorders (which are sometimes referred to in the medical literature as 'Pervasive Developmental Disorders'). It is common for the diagnosis to be undertaken by a multi-disciplinary team. No condition has been to the forefront of Irish news more than autism. Dozens of cases have gone before the High Court and at the time of writing, there have been several high profile cases of children being removed from the family home (these are discussed more fully from page 207). The education of children with autism is recognised as an area in critical need of continued improvement in Ireland.

The entire issue of securing appropriate services for these children is something of a battleground, fraught with difficulties, obstacles and controversy but conditions have improved quite a bit in the past six years. Minister Hanafin (Minister for Education and Science) recently stated that there are one hundred and forty one special classes for children with autism, ten pre-school classes and eight autism facilities have been approved by her Department. This is significant movement and must be applauded. However, educational resources for children with autism are still largely inadequate and a great deal more needs to be done.

What is autism?

Autism is a unique neurobiological condition that results in what is called a 'triad of impairment' which is defined as follows:

1. **Social interaction**:
   - Aloofness/indifference to others
   - Acceptance/enjoyment of social contact but inability to initiate/maintain it
   - Inappropriate/odd approaches to others
   - Paying little attention to responses given
   - Stilted/overly formal interaction

2. **Language and communication**:
   - Wide variation in communicative ability
   - Even extensive vocabulary can't be used correctly
   - Difficulty with social aspects of language: turn-taking, timing
   - Impaired social communication
   - Inability to read body language/gestures/pragmatic language
   - Inability to empathise
   - Literal understanding of language

3. **Imaginative thinking/range of activity/restricted behavioural pattern**:
   - Repetitive play activities
   - Unusual manner of using play materials
   - Preoccupation with irrelevant details
   - Stereotypical behaviour
   - Narrow range of interests/obsessions
   - Difficulty with planning and organising
   - Difficulty with change of routine
   - Problems with cause/effect
   - Difficulty with consequences of actions
   - Blurred distinction between fantasy/reality

The triad of impairment results in the individual with autism not perceiving the world as most other people do. Whilst every person with autism is different (because it is a highly variable condition), each has impairment in these three areas to one degree or another. As a result, children and adolescents with autism learn in both qualitatively and quantitatively different ways than other children. A significant number of children with autism also have an associated general learning disability within the mild to moderate range. Assessment of intelligence in children with autism is difficult; the assessment must be completed in a number of contexts with multiple observations in each and without over-reliance on the results of intellectual assessment (more on assessments from page 5).

Some children with autism will be able to receive their education in the mainstream classroom; others will require special classes or other settings. The more severe autism is often associated with a variety of difficult behaviours that will require focused intervention of a specific kind. Educational interventions for children with autism fall into a number of known types; Applied Behaviour Analysis, TEEACH, Gentle Teaching, and a number of others. Some methods require classroom and school settings that are quite unlike the ordinary school. What is certain is that a child with autism will require a distinctive form of education which recognises and accommodates their reliance upon visual cues and visual schedules, in order to perceive the world as an orderly, predictable place.

Children with **Asperger's Syndrome** have a form of autism that makes it more likely that they can benefit from education in a mainstream setting. Their level of intelligence is often within the average to well-above-average range, but their focus of interests can be restricted to one or two subjects (about which they can be quite obsessed and have deep knowledge). Like children with other forms of autism, children with Asperger's Syndrome do not usually develop close friendships, are not good at understanding their feelings or the feelings of others and find it difficult to sustain close partnership during cooperative learning tasks. They seem to be aloof and distant but they don't display many of the usual characteristics of other children with autism, such as the repetitive stereotyped movements, the unusual vocalisations or the disorganised and sometimes self-injurious behaviour.

The educational management and treatment of autism must begin immediately after diagnosis (a diagnosis is usually made between the ages of three and five). Education planning must proceed after careful assessment by the following: a psychologist, a psychiatrist, a speech and language therapist, an occupational therapist and sometimes a physiotherapist. Each of these assessments must yield information that will be translated into strategies and interventions written into the educational plan. It will probably be essential for these professionals to work directly with the child from time to time, often on a regular basis, over a period of many years.

Recommendations from occupational and physiotherapists are often focused enough to enable them to be implemented in school, by the teacher or aide. These recommendations need to be read carefully by all involved, with a view to discovering how and when they can be implemented. The exceptional difficulty parents have in Ireland in accessing these services at an early age (and beyond) is a significant complicating factor in the education of children with autism. Without early and appropriate intervention, and the consistency of that intervention over a number of years, the long-term outcome tends to be poor. The education of children with autism in Ireland remains the most significant weakness in the special education system.

One thing about the education of children with autism that must be included in the educational plan is a **home-school log**, or journal. In it, the class teacher and any specialist teacher, along with the special needs assistant, should make daily entries about the child's behaviour, learning and engagement in activities. The notes need not be long but they must be specific. Writing in a journal that "Brian had a good day today" conveys no useful information to anyone about Brian. Writing that "Brian made eye contact with Sheila for the first time today and pointed to the toilet while taking her hand" tells us something quite useful about Brian's progress. In the log teachers should also record what materials are being used with the child that week, how often and in what manner. The log should also clearly state whether or not parents are to supplement learning by working with the child at home, how, when and how often. If there is no home-school log, there is an important absence of information.

Specific Learning Disability

Children with specific learning disability (SLD) have been individually assessed by a psychologist and found to have a level of intelligence within the average range, 90-110 standard score points (25th-75th percentile). Despite their average level of intelligence, their basic reading, mathematics or writing skills fall at or below the 2nd percentile. The specific learning disabilities have several different names, some of which are in common parlance:

Dyslexia – difficulty with writing, reading and spelling
Dyscalculia – difficulty with mathematics
Dysgraphia – difficulty with handwriting
Dyspraxia – difficulty with motor coordination

The cause of each of these conditions lies within the pathways of the brain. There are early warning signs in the young child, which might include things such as:

Academic Symptoms
- Poor performance on group tests (Drumcondra, Micra T)
- Reversals in reading and writing
- Difficulty in copying accurately from a model
- Slowness in completing work
- Easily confused by instructions

Cognitive Symptoms
- Difficulty discriminating size, shape, colour
- Difficulty with the concept of time
- Distorted concept of body image

- Poor organisational skills
- Difficulty with abstract reasoning and problem-solving
- Disorganised thinking
- Obsession with one topic or idea
- Poor short-term or long-term memory
- Lags in development milestones (language, motor, etc.)

Physical Symptoms
- General awkwardness
- Poor visual-motor coordination
- Hyperactivity
- Overly distractible; difficulty concentrating
- Lack of hand preference or mixed dominance

Behaviour/Social Symptoms
- Impulsive behaviour; lack of reflective thought before acting
- Low tolerance for frustration
- Excessive movement during sleep
- Poor peer relationships
- Overly excitable during group play
- Poor social judgment
- Inappropriate, unselective and excessive display of affection
- Behaviour often inappropriate for situation
- Failure to see consequences of actions
- Overly gullible, easily led by peers
- Excessive variation in mood and responsiveness
- Poor adjustment to environmental changes
- Difficulty making decisions

The specific learning disabilities typically affect five general areas:
- **Spoken language** – delays, disorders and deviations in listening and speaking
- **Written language** – difficulties with reading, spelling and writing
- **Arithmetic** – difficulty performing calculations and operations
- **Reasoning** – difficulty organising and integrating thoughts

- **Memory** – difficulty remembering information and instructions
Children with specific learning disabilities typically receive their special education service from the learning support teacher, who will work closely with the classroom teacher. One problem frequently encountered in providing education for these children is the necessity of withdrawing them from the classroom for small-group tutoring with the specialist teacher. This often removes them from whole-class activity, such as reading or maths instruction, while they receive specialist attention in these key areas. The coming and going from the classroom is disruptive for the child and the class teacher.

There is a movement at primary school to provide supportive services in the mainstream classroom as much as is possible. This allows the learning support teacher to work with a group of children within the classroom with minimal disruption, and facilitates greater collaboration between the learning support and classroom teacher.

As with all assessment, assessment of specific learning disabilities requires the assessment of the general level of intelligence (psychological assessment) and the assessment of the academic areas that are of concern. The psychological assessment does not have to be repeated on an annual basis but the academic assessment **must** be repeated at least once a year in order to evaluate the success of the educational plan.

Specific Speech and Language Disorder

Children with a specific speech and language disorder (SSLD) have been assessed by a psychologist and while their non-verbal skills (Performance Scale) are within the average range, their verbal skills (Verbal Scale) are within the below-average range. These children have also been assessed by a speech and language therapist who has found that one or more of the main areas of speech or language are developed well below their non-verbal intelligence. These children do not have hearing, emotional or physical difficulties that are causing their speech/language difficulties. Children with speech and language *delays*, which may disappear over time, are not to be included in this category.

Children with specific speech and language disorders may escape notice because of the limited availability of speech and language therapy assessment in Ireland. Even if they have been assessed, they may have difficulty accessing speech therapy. Speech and language disorders do not just affect spoken language, they can interfere with the child's ability to understand what is said to him or her – there is also a difficulty in committing information into memory so it can be enacted upon. Therapy is always an important aspect of their educational programme and although it may be hard to obtain, it is possible for a speech therapist to provide guidelines, instructions and particular strategies of intervention for teachers and parents to implement in class and at home. One should not give up and assume a child cannot make progress unless professional therapeutic intervention is available.

Children with Special Education Needs Arising from an Assessed Syndrome

There are hundreds, if not thousands of syndromes that can have a negative impact on a child's learning or behaviour. Some of the more well known are Tourette's Syndrome, Down Syndrome, and William's Syndrome. Any syndrome that has been assessed and found to have a negative impact on a child's school performance will make the child eligible for special education services. The intensity and type of support and resources the child receives will always depend on the nature and profile of the child's strengths and weaknesses. Children with these sorts of conditions might receive services from a learning support teacher, a resource teacher and a language support teacher; they may have a special needs assistant provided to them as well.

Multiple Disabilities

This category of special need is reserved for those children who have two or more of the above conditions occurring at the same time. There are several special education conditions that co-exist at the same time. Children with autism often have a Moderate General Learning Disability as well. Similarly, children with ADHD may also have a Specific Learning Disability. When children present with Multiple Disabilities there are specific challenges to be met in educational planning. Each co-existing condition must be addressed simultaneously. It will do no good to focus on the Autism and ignore the General Learning Disability or to plan a programme for Giftedness and ignore a Specific Learning Disability if one is present. Educational planning of this type is complicated and requires scrupulous assessment of the impact of all the existing conditions on every aspect of the child's functioning. Following this rigorous assessment, an IEP must be drawn up which contains aims, goals and objectives for each condition. The level of support required may be quite intensive and involve multiple professionals across a number of disciplines.

Exceptional Ability/Giftedness

Giftedness is recognised as a 'disability' or special education condition in the Education Act of 1998. Presently there is some disagreement about the type of services these children are eligible to receive. The focus in special needs tends to be on the more extreme end of the difficulty continuum and Gifted Children are often not perceived as needing special education. This is unfortunate because it is well known that Giftedness carries with it a need for special services.

Children who are considered gifted have been assessed by a psychologist and found to have a high level of intelligence (an IQ score of 130, 98th percentile). This is the most simplistic definition of giftedness and one that is not particularly helpful or informative. According to the Special Education Review Committee, whose report was released in Dublin in 1993, giftedness can be defined as:

a) General intellectual ability
b) Specific academic aptitude
c) Creative or productive thinking
d) Leadership ability
e) Visual and Performing arts skills
f) Mechanical aptitude
g) Psychomotor ability e.g. in athletics, gymnastics

This model can be augmented with an older model, provided in 1957 by DeHaan and Havighurst, who proposed that 'giftedness' could be viewed as comprising of six domains of excellence (a child could be 'gifted' in one domain, or in several) as follows:

- **Intellectual ability:** relates most directly to school subjects and encompasses the verbal, numerical, spatial, memory and reasoning factors of the primary mental abilities. DeHaan and Havighurst regard a combination of these aptitudes as being basic requirements for other talents such as fine arts, social leadership, science and mechanics.
- **Creative thinking:** the ability to recognise problems, to be flexible in thinking, to think laterally, originate ideas or products or to find new use for old objects or materials.
- **Scientific thinking:** numeracy skills, the understanding and use of algebraic concepts, arithmetic reasoning, curiosity about the natural world and facility with scientific methodology.
- **Social Leadership:** specifically the ability to help a group to reach its goals and to improve human relationships within a group.
- **Mechanical Skills:** otherwise known as 'craft skills', they are closely related to talents in the fine arts and sciences. Successful intelligence in this category depends on manipulative facility, spatial ability and perception of visual patterns, details, similarities and differences.
- **Talents in the fine arts:** artistic, musical, verbal and written skills.

Whatever the model of giftedness and whatever the terminology, ('gifted' or 'exceptionally able'), it is now clear that children who fit in this category have learning needs which are just as genuine and critical as those whose special education conditions place them lower on the scale of intelligence or academic progress. There is a common misconception that children who are gifted will somehow educate themselves, that all they require are advanced textbooks or placement in higher classes for them to thrive. Most of the focus in special education is at the lower extremes of achievement and it has resulted in a nearly universal lack of acknowledgment that children who are gifted also require special services and written educational plans.

There is another difficulty related to children who are gifted. Because their ability is so high in so many areas, there is a misconception that they cannot also have specific learning disabilities, or ADHD. The fact is that a child can be both gifted and have an autistic spectrum disorder as well. This combination of special needs requires careful programming if the child is to succeed in school. Gifted children who have other special education conditions, have been referred to as 'twice exceptional students'. One author has identified three possible subgroups of children who may be both gifted and have other special needs, as follows:

Group 1 – children identified as gifted but underachieving in school

Group 2 – children with recognised learning difficulties but whose giftedness has gone undetected

Group 3 – children considered unqualified for services for giftedness or for other special needs because their abilities and disabilities mask one another (this may be the largest of the three groups)

Some early signs of potential giftedness may include:

- Advanced progression through developmental milestones such as talking, crawling, etc
- Curiosity
- Early and extensive language development
- Early recognition of caretakers
- Enjoyment and speed of learning
- Extraordinary memory
- Less need for sleep in infancy
- Sensitivity and compassion
- Perfectionism
- Unusual alertness in infancy

Educating Gifted Children

Because the needs of the gifted in Irish schools often go unrecognised it may be helpful to draw the reader's attention to some simple, basic facts a teacher can keep in mind when differentiating instruction for children in the class who are advanced learner's, whether gifted or not. Tomlinson, writing a digest for the Educational Resources Information Centre (ERIC) has listed the following points:

- "Concrete to abstract. Learners advanced in a subject often benefit from tasks that involve more abstract materials, representations, ideas, or applications than less advanced peers.
- Simple to complex. Learners advanced in a subject often benefit from tasks that are more complex in resources, research, issues, problems, skills, or goals than less advanced peers.
- Basic to transformational. Learners advanced in a subject often benefit from tasks that require greater transformation or manipulation of information, ideas, materials, or applications than less advanced peers.
- Fewer facets to multi-facets. Learners advanced in a subject often benefit from tasks that have more facets or parts in their directions, connections within or across subjects, or planning and execution than less advanced peers.
- Smaller leaps to greater leaps. Learners advanced in a subject often benefit from tasks that require greater mental leaps in insight, application, or transfer than less advanced peers.
- More structured to more open. Learners advanced in a subject often benefit from tasks that are more open in regard to solutions, decisions, and approaches than less advanced peers.
- Less independence to greater independence. Learners advanced in a subject often benefit from greater independence in planning, designing, and self-monitoring than less advanced peers.
- Quicker to slower. Learners advanced in a subject will sometimes benefit from rapid movement through prescribed materials and tasks. At other times, they may require a greater amount of time with a given study than less advanced peers so that they may explore the topic in greater depth and/or breadth."

It is difficult to generalise about gifted children because their characteristics are so individual. However, following Tomlinson's guidelines a teacher interested in teaching to the skills of the most able children in the class will do some of the following things that are central to differentiating instruction.

Modifying the Content

Children should move through content at their own pace, however fast (or slow) that may be. If a theme or skill is mastered, they should be provided with more advanced activities, not more of the same activity. Integrated content is more useful for the gifted than single-subject teaching. Concept-based teaching helps these children generalise and apply ideas in new, creative ways. Broad-based, thematic content is most suitable to the learning style of the gifted. These children also benefit greatly from integrated teaching that goes across unexpected boundaries, for example learning about Irish history while incorporating the study of dances, ancient sports, and party games.

Modifying the Process

Activities need to be restructured to become more intellectually demanding. Challenging gifted children with questions that demand high level thinking and creativity is one way to increase the intellectual load of a task. Having them generate their own questions and then engaging in research to find the answers is another. Active exploration and discovery learning are the key strategies for the gifted. The goal is to get them to think about the subject in more abstract and complex ways. Self-directed learning should be encouraged and based as much as possible on the interests of the child. Analysis, synthesis and evaluation of content are to be both encouraged and demanded. Guided self-management, group interactions and flexible pacing are a few of the methods helpful in managing class activities.

Modifying the Environment

A learning environment that includes a wide variety of materials, print, audio, video, and computer is most appropriate for gifted children. They should be allowed to move as much as possible from one learning centre to another. The teacher should encourage independence and the development of novel solutions to problems. Facilitating and supporting learning outside of the classroom is also helpful. Advanced resources must be provided in the classroom whenever possible. It may be necessary to provide college level textbooks for the most advanced learners in the school. Every effort must be made to provide stimulation, generate inquiry, and stimulate creativity. It is a myth that gifted children will advance on their own without adult guidance. It is the responsibility of the teacher to support the learning style of the gifted with resources and a stimulating classroom environment.

Parenting Gifted Children

Parenting a gifted child brings with it complications that are unique to the child, the family, and the circumstances in which they live. No two gifted children are alike so it is difficult to provide specific advice in a book of this scope. I want to focus on one particular aspect of giftedness in relationship to parenting though. That aspect is what is called "**dyssynchronous development**". Dyssynchronous development refers to the developmental characteristics of gifted children in which their intellectual development vastly exceeds other areas of development. Social-emotional skills may be age-appropriate as may physical skills. Because gifted children so often appear sophisticated beyond their years (see the section about the three ages of the child on page 74), adults often over-estimate their abilities to get along with other children, to fit in with a group, and to understand the subtle nuances of social interactions. Also, parents can become obsessed with academic skills and cognitive development to the exclusion of social-emotional development. Learning to get along with others and cooperate with group activities are important skills that must be learned in childhood. The parents of a gifted child must recognise that these skills provide the foundation upon which the intellectual and cognitive gifts of the child will rest. I urge parents all parents, but especially parents of gifted children, to recognise the whole child before them, not just the head and what is in it.

A Note About the Category of Gifted/Exceptional Ability

The reader's attention needs to be drawn to the contradiction in the inclusion of Giftedness as a special education condition between the Education Act of 1998 and the Education of Persons with Special Education

Needs Act of 2004. The category of Gifted/Exceptionally Able is included in the Education Act but not in the ESPEN Act. The implications of this are unclear at present. Whether or not the DES intends to extend the entitlement to special education resources to Gifted children will become apparent in the next year or two.

PART THREE

Let's understand more about Special Education

1. Understanding children with Special Education Needs

Children with special education needs present challenges that other children do not. Most of the challenges they present are a result of flaws in *our* vision and understanding. When education and child-rearing are thought of as a set of one-size-fits-all strategies, we lose sight of the differences in children. There is no one way to learn, no one way to behave and there is no one solution that will solve all children's learning or behavioural difficulties. What is needed on the part of adults is understanding; understanding the meaning and background of a child's behaviour and understanding the lives they have lived/are living.

We must learn to recognise and understand the struggle that a child with special education needs faces, on a daily basis, in and out of school. Even the mildest special education condition results in big challenges in learning and living. As previously stated, it is now recognised that most of the common special education conditions and learning difficulties (such as dyslexia, ADHD, and behavioural problems) result from neurobiological irregularities in the child's brain. Something is going awry with the electrical-chemical activities in the brain. Children do not *choose* to have learning difficulties and they do not choose to have behavioural problems. Different is just different and difference in learning or behaving requires difference in teaching and parenting.

When a teacher has the philosophy of "I teach the same way to all the children and they will just have to adapt", or parents have a philosophy of "I raised all my other children this way – he'll just have to adjust", they miss the essence of the child and create a world of difficulty for children. We are not all alike; we are not all born with the same gifts and talents. The famous psychologist and contemporary of Freud, Alfred Adler, said long ago, "the world has need of many excellences. There is poverty within us when we fail to recognise the gifts a child brings to the classroom."

For some children, just getting into school each day is a triumph of the will. Children with specific learning difficulties sit in a classroom all day long and watch other children read more fluently than they can, write more easily, do maths more quickly. The other children watch *them* struggle to learn. Imagine the courage it takes to arrive at school knowing you might be called on to answer a question you can't answer, to read aloud when you can't make sense of the words. The amount of discouragement these children face day in and day out is staggering. Yet they come to school, want to learn, want to please the teacher and make their parents proud. We must never forget how hard it is to face these obstacles every day.

Discouragement is the enemy of academic progress and behavioural improvement. The discouraged child, sooner or later, will try to establish his or her belonging, one way or another. Often they begin to pursue their belonging in troublesome ways. Some become the class clown, seeking attention with little provocations. Some declare learning bankruptcy, shut down and produce little work. Others become angry or withdrawn.

Having a learning disability is not just a learning problem; it is a problem of living and fitting in with your peers. A significant number of children with specific learning disabilities have social problems secondary to their learning disability. They are sometimes not particularly good at interpreting facial expressions; tone of voice, or in understanding humour, sarcasm or irony. Difficulties of this sort, if they occur along with a learning disability, will always result in social problems for the child. Everyone wants to fit in but for some it is just hard to find the door.

Children with learning difficulties and special education needs often have so many failure experiences in school that it is a wonder they want to go to school at all. Day in and day out, teacher has to correct their mistakes (why can't we focus on success and improvement more). Spelling errors are corrected, wrong answers pointed out in front of peers. This is not a happy way to live your life as a child. When successful learning experiences are few and far between, a child can easily become de-motivated and discouraged. A de-motivated and discouraged child cannot learn efficiently. What began as a cognitive problem secondary to a special education condition is now a psychological problem.

Discouragement leads to lower self-esteem that leads to lower self-confidence and fewer efforts to try and succeed. This vicious cycle of failure disrupts the learning cycle. There are a great many adults among us who are victims of this cycle and who left school early feeling as though they were a failure.

Children with **ADHD or emotional/behaviour problems** are the single most misunderstood group of children in school and society today. They are blamed by everyone for having a condition they did not choose to have. Nearly every day of their lives they are scolded, corrected, shamed, punished and blamed. How do you live this way and not become bitter and cynical about adults and the world around you? Trying to navigate your way through a world you don't understand and which rejects you takes a toll on one's sense of self. It is time to recognise that it is not their fault. I have never met a child who wanted to be punished and scolded every day! Any teacher or principal will tell you that they spend 80% of their time disciplining the same 20% of the students in a school. We find it difficult to consider that the problem may lie in *our* solutions, which are inadequate to correct the errors in living that these children make. It is no wonder that children with significant ADHD emotional, or behavioural problems often become depressed, leave school early, and drift into the more useless side of life and living. Who wants to be reprimanded every day, reminded that you are bad, sent out of the classroom and rejected by your peers?

Social Skills

Schooling is about more than just acquiring academic skills. It is not enough to be a brilliant mathematician if you cannot understand yourself or others, cannot cooperate on group projects, cannot contribute to society because you are hell-bent on getting your own way. We are social animals, we live in groups, and we get our sense of belonging from our relationships with other people. When education focusses on academic instruction, to the neglect of social and emotional development, we are in danger of educating a group of bright and talented misfits.

The social and emotional development of children in school and in the home is the most important thing we must focus on. The ability to cooperate and contribute are the two keys to a successful career, no matter what career path is chosen. Children with special education needs, no matter what the condition, must have opportunities every day in school and at home to make a contribution to group projects at the level of their ability and regardless of the behavioural challenges they bring to the classroom or family. Everyone wants to contribute and sometimes the seemingly most challenging behaviour is only a mistaken pathway to contribution.

How do children learn? They learn by making mistakes, the same way adults learn. The only bad mistakes are the ones you don't learn from. If children learn by making mistakes, then it is only reasonable to expect them to make mistakes in living. Few of us master a task on the first try. The more complex the task, the more likely it is we will need several chances to achieve mastery. No tasks in living are more complicated than the task of social interaction. The world is filled with many different people with different attitudes, ideas, beliefs and values. School and family life needs to be a place where children can learn, step by step, to interpret their environment and understand the needs and motivations of the people around them. There is no shortcut to that sort of learning. Because children with special education needs often have more difficulty than others interpreting the world around them, they will need more time, more chances, more practice and more patience to succeed.

Continuum of Provision

In special education there is increasing recognition that there must be a continuum of provision for children i.e. a long-term individual plan. No two children with the same condition are alike – each has a different profile of skills and abilities. Because of this, all special education must be tailored to the individual needs of the child. No one programme can be created that will meet the needs of all children. No one placement can suit all children. Some children with special needs will benefit from low level interventions in the mainstream classroom; some will require more intensive interventions in specialised settings. What is important is that the particular type of special education provision is created with the child in mind. Educational programmes for children with special education needs are made to order. The starting point and ending point is the child, not the system. Appropriate special education provision is child-centred, not system-centred.

Punishment

Why do we punish children when they do wrong? Is it because we were often punished when we were children? Is it because our teachers spent so much time working out various 'sanctions' and 'discipline policies' in our schools? Here's a startling fact: punishment is ineffective in teaching children how to live a useful life! Punishment *temporarily stops* behaviour, but does not *teach* a child how to behave. The child who lives with too much punishment will become a punishing adult. Punishment comes in many forms; shame and humiliation at home or in school, removal from the group and isolation in a room or school office, restriction from pleasurable activities and so on. For some children the world is just a punitive place they must endure.

The issue of corporal punishment has taken much space in educational literature over the decades. We know there are many adults who believe that children should be smacked, who believe that corporal punishment is the best way to correct children's wrongdoing. It might be helpful to provide an 'English translation' to the term 'corporal punishment'. Corporal punishment is the act of an adult deliberately inflicting pain on the body of a child. When we think of it this way, instead of the more antiseptic understanding, we can see what it really is. Corporal punishment teaches children that adults are pain-inflicting people who will hurt them if they make a mistake. All children want to learn the right way to live. Punishment in any form, particularly corporal punishment, will not teach them how to live the right way.

Part Three

2. The Essential Elements of Special Education

Mainstreaming, Integration, Inclusion and Segregation

The inclusion debate isn't much of a debate any longer. There is a worldwide trend in attempting to include children with special education needs in mainstream classrooms. Some countries have included nearly all children; some include none. The inclusion debate is as much a social debate as it is an educational debate. Is there any solid, conclusive evidence that children with special education needs benefit from being included in mainstream classroom as opposed to receiving their education in a separate, specialised setting? The honest answer is no, there is no conclusive evidence. But the answer to the question depends largely on what outcome measure one is looking for. If we are looking for evidence of academic achievement, then we have to say that there is evidence on both sides, children learn well in some inclusive settings and in some segregated settings. If we are looking for social evidence, then there seems to be more success if children are included. We might benefit from a clearer definition of the terminology.

Mainstreaming

Mainstreaming was one of the first developments in special education. Instead of sending children to special schools, children were enrolled in special classes in 'ordinary' schools. While enrolled in these classes, the children sometimes participated in lessons or activities with their same-age peers in a regular classroom. For example a child might be receiving his education in a special class but visit the regular classroom during music instruction. Mainstreaming was an effort to stop segregating children from their same-age peers. Because all of the children appeared to benefit from this type of provision, efforts were made to increase opportunities for children with special education needs to be educated with their peers who did not have special needs. Integration was the next step on the ladder to inclusion.

Integration

Integration of children with special education needs differs from mainstreaming in significant ways. Instead of *occasionally* receiving lessons with other children, the child with special needs who is integrated is enrolled in the same class as other children, but may leave the classroom from time to time to receive specialist assistance. Children who are integrated may receive most of their education in the mainstream classroom but will also receive specialist services in separate rooms from specialist teachers.

Inclusion

Inclusion is the placement of a child with a special education condition in a mainstream classroom with all the necessary support and resources in place. Supports range from low-intensity such as preferential seating placement, or differentiated assignments, to high-intensity such as nursing care, assistance in feeding and toileting. Successful inclusion is dependent on the provision of resources for all included children. When resources are in place children can be successfully included. At the present time in Ireland there is a critical lack of resources and the various pieces of education legislation make it clear there is no real intent to provide resources based on need. Meaningful inclusion will be difficult under these circumstances but most children with a special education need can be included in mainstream classrooms, with the required specialist services brought to them.

Segregation

Ireland has about one hundred and thirty special schools at the present time. There is a long history of special schools in Ireland. These schools have been centres of excellence in the provision of special education and have a history of successful education for a great many children. With the social movement towards integration and inclusion, there is declining enrolment in the special school.

The social ethos behind the movement towards integration and inclusion is a powerful one; children who are segregated from their peers often end up entering a world of adults who do not understand them and have difficulty accepting them. The tension between special schooling and inclusion creates difficulties for all concerned. Teachers in special schools most often have sets of highly specialised skills and knowledge that would benefit all teachers. As pressure mounts on the special schools as a result of declining enrolment, they will need to adapt and change. It will benefit all if they continue to be centres of excellence and create opportunities to share their expertise with teachers in mainstream schools.

Restrictiveness of Environment

The principle of least restrictive environment (LRE) is central to appropriate special education provision. The LRE principle means that the setting in which a child receives special education will allow for the maximum interaction with other children, given the nature and severity of the condition. All special education provision proceeds along the LRE scale, with changes made only after every possible intervention has been attempted in that level of restriction. The LRE is a range that goes from least restrictive to most restrictive:

1. Mainstream classroom with classroom teacher modifications
2. Mainstream classroom with specialist support provided in the classroom
3. Mainstream classroom with specialist support provided occasionally in a different room in the same school
4. Special class for some subjects, mainstream class for all other subjects
5. Special class for most subjects, mainstream class for some subjects
6. Special class for all subjects
7. Special school
8. Special day facility
9. Residential facility

3. The Role of Play in Special Education

"Play is the work of childhood." So said Friedrich Froebel, 19th century educational philosopher and founder of the Kindergarten. Play was central to Froebel's method of teaching children – for Froebel there was no possibility of education *without* play. We have come a long way since Froebel founded his Kindergarten and even though his movement and philosophy swept Europe and America, we have lost the essence of his message that 'play is central to the education of children'.

Why is play so important? It is the natural propensity of the child to play; to run through puddles, roll down verges, build sand castles on the strand and turn over rocks to look for insects. This natural tendency of children to engage in play, to be imaginative, to use familiar objects in unfamiliar ways is present for an important reason: the brain grows and develops through play. Let's take a look at how this happens and why it is important for children with special needs to play.

A child climbs up the ladder to a slide. Slowly, hand over hand, one foot on a rung, then the other foot on the same rung, to the top she goes. A bit afraid to stand, not sure if she can sit, finally she manages to negotiate herself at the top of the slide, seated and ready to go. The ground looks far away. A bit nervous about what might happen next she pushes off, rather gingerly, but pushes off just the same. Down she goes, faster and faster, the end of the slide coming closer and closer. What will happen next? Boom! Right on the sand, bottom first; it hurts a little but not too much. Laughing and giggling, looking around to see if Mum saw her accomplish this momentous task, she runs to the ladder and starts to climb again. Let's look at what's going on inside the brain?

First of all there is the sighting of the tall ladder and the seat at the top. Her brain is learning about higher and lower. Then there is the climbing up the ladder. Her brain is learning about weight, density and mass as she tries to pull herself up each rung. Anxiety accompanies this climb and her brain is learning to control anxiety by thinking that things will be OK … Mum isn't far away. Her loving attachment to mammy helps her go up the slide, Mum is a person I can trust, she will be there for me if I need her. One foot at a time, two feet on each rung before the foot rises again to the next rung. The brain is experiencing balance and coordination and left-right orientation. At the top of the slide she can see the ground below. Her brain is experiencing higher and lower, above and below. Now going down the slide, the brain experiencing the forces of gravity. Slow at first, then faster; the brain experiencing inertia and velocity. Off the end of the slide with a bump, landing on her bottom. The brain experiencing that physical, moving objects want to keep moving; density, mass, gravity again.

The physical experience helps the brain to programme the neural pathways to a state of awareness that will facilitate the learning of abstract, non-physical concepts much later in life. The more we deprive children of opportunities to play, the more we deprive the brain of opportunities to learn. Children who lack play experiences are at greater risk of learning difficulties in primary school and even in secondary school. When mathematical concepts are acquired through active learning experiences, even if mathematical language is never used in these experiences, it is much easier to master the maths curriculum at primary and secondary level. Play is the royal road to acquiring basic skills in self-control, self-confidence, and self-esteem and becoming a creative, capable adult who lives life with a sense of fun and a playful outlook.

Emotional and social skills are also acquired through play. When children play pretend games, when pencils become airplanes and markers become rockets, experiences are shared and the creative imagination is stimulated. It is through play that we learn to share, take turns, wait and watch, tell stories, control our impulses, tolerate frustration, develop language and social communication skills and extend our minds into the realm of the unknown and unpredictable. Play is a place where we can face our fears in safety and know that the imaginary terrors we encounter can be overcome and defeated. Play is a place where the emotional centres of the brain learn control and modulation. Play is also a place where empathy and compassion for others is built and solidified. Anyone who has observed children at play has seen one child start to cry and another crying shortly after seeing this. Our capacity for empathy is one of the noblest attributes of the human being.

Children with special education needs are in need of playful and imaginative learning experiences. For all children, but especially for these children, play should be an integral part of every school day. Play is education in activity and is important to the educational life of the child. When a school curriculum downplays the role of play it creates educational disadvantage for children and when this happens to children with special needs they are at greater risk of learning difficulties than others. It is imperative that we take a look at educational planning for children with special needs and build into these plans opportunities for playful, imaginative experience. An obsessive focus on reading and maths is not helpful. All educational plans for children with special needs should contain statements about how the child will be supported in their playful activities, what steps will be taken to improve their play, and when and where this will happen.

4. The maximisation of scarce resources and services

Children with special education needs will often require specialist supportive services as part of their education. For some children it is essential they have speech and language therapy and assessment at an early age (children with autistic spectrum disorders for example). Accessing these services can be problematic, even impossible at times. The shortage of speech and language therapy services for children with autistic spectrum disorders borders on gross negligence. It is self-evident in special education that the earlier the essential services are provided, the better the outcome.

It is a reality however that many supportive services are not readily available. What can we do to make up for the shortfall? The classroom teacher and specialist teacher can go a long way to fill the gap. There are a lot of activities that can be built into the classroom week, or into the schedule of work of the special education teacher, to supplement what services need to be provided outside of school.

For example, an Occupational Therapist often works with children who have motor coordination, balance or motor control problems. Following an OT assessment, recommendations are usually made for a variety of activities that will help the child. These often involve balancing, catching, throwing,

Part Three

crawling in, under, and around objects, etc. These activities, if they are provided in written form to the class teacher, can easily be incorporated into the PE programme, since *all* children enjoy and benefit from these activities. They can also be provided to the special education teacher who will engage in them with the child, during some of her sessions.

Likewise, speech therapists sometimes work with groups of children to develop language skills. Although speech therapy, like OT, is a highly specialised profession, there is no reason that some helpful suggestions can't be provided to class and special teachers who can incorporate some of them in circle time activities or small group work. Language development does not depend entirely on the presence of a speech therapist. There is a lot a class teacher can do to stimulate language development. Statements to young children such as, "Use your words", "tell her what you want", or "Thank you for using your words" direct them to the importance of talk and language.

The point is that we as adults need to think outside the box and recognise that if supportive services are not readily available, we have to do something to assist. When we have developed this mindset we will open possibilities for all children with special needs to grow and develop in school, at the same time as the curriculum is expanding to include educating the whole child, not just the head.

5. Whole School Planning

When it comes to special education in mainstream schools nothing is more important than whole school planning and no one is more important than the principal. When the principal has an open attitude to including children with special education needs, he or she brings the entire staff along with them on that journey. Children with special needs are not the sole province of special education teachers or supportive service providers such as psychologists. Once a child is accepted in a school, the school assumes responsibility to meet the needs of that child to the best of its ability.

The Department of Education and Science now requires that schools have written policies about admission and support of children with special needs. Each school must have a designated special education support team that will oversee its provision in the school and serve as a support 'service' for teachers, parents and children. The support team will work in consultation with the Special Education Needs Organiser (SENO, see page 189) assigned to the school and assure that the maximum effort is being put into educating all children, especially those with special needs. There can be no real improvement to the provision of special education in a school unless there is 'whole school planning' to support it.

Whole school planning and decision making requires concerted effort on the part of the entire staff. Attitudes must be explored in an atmosphere of openness and without judgment. Some teachers find it difficult to consider that children with special needs will be placed in mainstream classrooms. There is no shame in this view. A school is an educational institution and an educational institution is one where ideas are shared and the truth is spoken, whatever that truth may be. Only if we begin to share ideas can they evolve, mature and blossom into new truth.

Under the leadership of a principal with vision, who is aware of the legislation and Departmental guidelines, provision can improve, strategies can be shared, solutions explored, and innovative programmes can be implemented, reviewed, revised, and refined. Parents might like to ask when the last staff discussion of special education issues was undertaken, what ideas were shared, and what new solutions have been put in place.

6. Pathways to developing Social Skills

Although this book is not meant to be a 'how-to' manual, I want to highlight the importance of a sequential programme of social skill development, from infant classrooms to senior classrooms and on to secondary school. As stated previously, nothing is more important to the development of the person than social competence, the understanding of self and others.

Social competence in infant classrooms begins with the development of a 'feeling vocabulary'. A feeling vocabulary consists of the words children use to describe their emotions, e.g. sad, happy, angry and so on. When children have words to label emotions, they reduce the need to act out the emotions. If they can use their words, they don't need to use their hands or feet. A feeling vocabulary in infant classrooms is developed in a playful, creative way. Masks can be made from paper plates, the children being asked to make a 'happy mask' or a 'sad mask' etc. Photographs cut from old magazines and newspapers, in which people display obvious emotional expressions, can be used by children for sorting activities (links to the maths curriculum). Cartoons from television can be videotaped, shown to the children with the volume turned off, the frame frozen and the children asked how they think the character is feeling. The drama curriculum can be used, the children asked to stand as if they were happy, stand as if they were sad, angry, etc. All of these activities draw the attention of children to the realm of emotions, the importance of words to talk about emotions, and the importance of interpreting facial expression and body language in understanding how others feel.

Children will make social mistakes; it is a natural part of growing up. Sometimes frustration tolerance is low and a child will push or hit. Young children should be asked, "Was that a friendly thing to do?" or, "What did I see you do with your foot just now?" The point in framing the question this way is that it avoids judgement and focuses on the behaviour. When we focus on behaviour we remove personality from the equation. When we ask a child if something was a friendly thing to do we communicate to them the need to establish friendship. Our guiding principle is that children will make mistakes in social interactions and that mistakes are things we need to learn from, and that we will look for progress over time, not instant results. There is no need to punish these minor episodes of social mistakes; what is necessary is a concerted programme of instruction to teach better social skills.

The Four-Question Technique
A useful technique for children who exhibit more serious social errors is this technique of proven effectiveness. Children are asked four questions,

at a time after things have cooled down and tempers have settled (including the teacher's/parent's):

1. What did you do?
2. What happened when you did it?
3. What could you have done differently?
4. What would have happened if you did it differently?

Let's take these questions one at a time and analyse them.

Question One: "What did you do?"
This basic question focuses on behaviour. It needs to be asked in a gentle but concerned manner. If the answer is "David made me...", you simply say to the child "I will be asking David what he did soon. Right now I want to know what you did." If asked in a calm and respectful manner, most children will answer honestly. I believe that most children who refuse to answer questions like this honestly have been taught by *adults* not to be honest, either by observing the adults lie and cover-up their own mistakes or by having been unreasonably punished after answering honestly.

Question Two: "What happened when you did...?"
This question is an attempt to link cause and effect in the child's mind. Some children do not make an automatic connection between their actions and the responses of others. Although it may seem obvious to us that they should know, they don't know. Therefore it is our responsibility to teach them. (Notice that all the questions are 'what?' questions, not 'why?' questions. There is often no answer to the why question. Sometimes we just don't know why we did the things we did. What we are looking for in the response to this question is recognition on the part of the child that something happened when they did what they did. An emotional reaction to the question, that focuses on the other child involved, is a positive sign. If we do not get such a response, we know we will have to work on the development of empathy and compassion and must build that into our dealings with that child.

Question Three: "What could you have done differently?"
This is a diagnostic question, diagnostic in the sense that the answer will inform us how we must proceed, over a period of time, to teach the

child new social competence skills. There are several possible responses to the question:

Example One: I should have told the teacher; I should have walked away. This response tells us the social skills of the child have capacity for pro-social alternatives (as opposed to anti-social). We therefore do not have to *teach* pro-social skills because we know they are in place, the child himself or herself has articulated them. When pro-social alternatives are present, what we need to do is foster and encourage their gradual re-emergence.

Example Two: I should have hit him in the face; I should have kicked him. This sort of response informs us the child has anti-social skills instead of pro-social skills. We should not accept at face value this sort of response as being entirely representative of the child's social competence. Some prompts such as "tell me something else you could have done instead?" may get a more pro-social response. In that case, we are back to example number one. However if we continue to get anti-social responses we know we have a lot of work ahead of us. The child must not only be taught pro-social skills but must also be enabled to eliminate the anti-social ones. This will take time and will be intensive work; it will require a coordinated and concerted effort from all involved with the child, at home, at school and in the community.

Example Three: The child says nothing or shrugs their shoulders. If the child cannot summon any response at all, and we have reason to believe that he or she genuinely does not know how what else to do, we have to teach new skills. They are not there to be encouraged out of the child; they must be put in their 'skills system'. Although it may take a while, it is not as difficult as in example number two, because there are no behaviours to be extinguished.

Question Four: "What would have happened if you did it differently?" Again this is another cause and effect question. The goal of social competence training is to constantly make efforts to link behaviours we exhibit with their impact on others.

USING THE FOUR-QUESTION TECHNIQUE

This technique is a teaching technique but it can also become a discipline technique. Having had this brief interchange with a child who has made a serious social error, we can require the child to write out the questions, (or we can give the child the questions on a photocopied sheet) and ask them to write down their response to each question. They must use their best handwriting while doing so. The child is given three sheets of paper. The child must write the answers on *each* sheet of paper. Three repetitions of writing is the maximum and the child's responses are written exactly, or nearly as exactly, as they have said them. This technique can even be used with children in infant classroom who can't write! I have seen it used this way. Young children are at pre-writing stage, in which they engage in scribble writing. This is a play activity that is a pre-cursor to real writing. Capitalising on this natural activity of children, a young child can be asked to write out their responses. While they are engaged in this scribble writing, they are mentally saying and repeating the responses to themselves. This serves as a calming and de-escalating activity.

The four-question technique is a powerful tool in helping children learn from their social mistakes. The theory underscoring this technique is that all behaviours are learned. Social mistakes arise because of faulty learning. It is based on the belief that children learn by making mistakes, that most social difficulties are a result of mistakes, and that as teachers and parents we can teach and encourage skills of social competence.

7. Understanding Challenging Behaviour*
a version of this article appeared in the newsletter of the Irish Primary Principal's Network

The issue of challenging behaviour is of increasing concern to educators at every level of schooling. In today's world, some children are coming to school with serious anxieties, a history of poor early experiences and familial difficulties. This can lead to a variety of behaviours that can disrupt the learning environment for themselves and others. A lot of work is going on to create and sustain interventions to reduce the frequency and severity of behavioural disturbances in schools.

For these interventions to be successful, an understanding of the psychological, social, familial and brain-related factors that contribute to challenging behaviour is the first step.

What is challenging behaviour?

Challenging behaviour is difficult to define. It is not a diagnosis and not a special education condition (although it can accompany several special education conditions). The educational literature does not contain a definition but the one featured in the INTO handbook is a good reference point:

> "Behaviour of such intensity, frequency and duration that the physical safety of the person or others is likely to be placed in serious jeopardy, or behaviour which is likely to seriously limit or delay access to, and use of ordinary facilities" (Emerson et. al. 1987) cited in INTO "Managing Challenging Behaviour."

Challenging behaviour takes a number of forms, and the INTO booklet describes it as anything which:

- Interferes with the pupil's own and/or other pupil's learning
- Challenges the day-to-day functioning of the school
- Challenges the right of staff and pupils to a safe and orderly environment
- Has a duration, frequency, intensity or persistence that is beyond the normal range of what schools tolerate
- Is less likely to be responsive to the usual range of interventions used by the school for misbehaviour (INTO, Managing Challenging Behaviour)

From the educational perspective, it is a type of behaviour most unlikely to respond to the customary strategies used in the classroom and school. Behaviour is challenging when our efforts as adults, (educators and/or parents), assuming they are appropriate in the fist instance, fail to reduce either its frequency or intensity.

What causes challenging behaviour?

Challenging behaviour, whether it occurs in children, adolescents, or adults can arise from a number of different causal factors that include, but are not limited to:

- Special education conditions
- Traumatic Brain Injury
- Schizophrenia, Bi-Polar Disorder
- Socio-economic Disadvantage
- Attention-seeking
- Communication difficulties
- Dysfunctional family systems
- Dysfunctional schools
- Dysfunctional teachers
- Developmentally inappropriate methodology
- Child temperament
- Educational neglect
- Abuse, trauma, chaos
- Senile Dementia
- Alzheimer's Disease
- Huntington's Disease

Given the fact that the cause of challenging behaviour can be varied, it is critical for educators to be mindful that whatever interventions are applied, be they at classroom level or school policy level, they must be tailored to the *cause*. For example, interventions for challenging behaviour that arises from ADHD, if applied to children with autism, will likely be harmful to the child and lead to increased difficulties. For this reason it is not possible to generate a one-size-fits-all intervention plan, or to find a manual of quick fixes. Thus, before anything is done to create interventions, it is necessary to investigate the causal factors.

Issues in Identifying Challenging Behaviour

There can be great variation in what is identified as challenging behaviour; a lot depends on the person who is identifying it. All behaviour is relative to a context – what is challenging in one context can be perceived as quite normal in another. There is a point at which a behaviour ceases to be simply 'irritating' and becomes 'challenging'. What criteria are used to make this judgment? It is well recognised in schools that a child who is described as challenging by one teacher is perceived as a typical youngster by another. All teachers, like all parents and all adults, have differing thresholds of tolerance for behavioural variations. We must exercise caution before we conclude that a child is exhibiting challenging behaviour. As hard as it may be to consider, there are times when the problem is within us, not the child.

The old question of nature or nurture has been answered definitively now. It is neither one nor the other but both; it is how our nature is nurtured that largely determines our behavioural repertoire. There are biological factors that put an individual at greater risk of developing challenging behaviour (a history of family mental health problems etc.). There are gender related issues too; from infancy, males are played with more vigorously than girls; males are allowed to engage in more active play and have behavioural patterns that are tolerated differently when they occur than if they occur in females. There is research that seems to indicate that the male sex hormone plays a role in aggressive behaviour in boys. A definitive answer to some of these gender issues has yet to be arrived at.

Ethical issues will always raise their head when attempting to create interventions, programmes and policies for children with challenging behaviour. What sorts of measures are appropriate? What is the role of punishment? Are sanctions appropriate? What behaviours will we attempt to change and what cost will the child pay if we are successful in changing them. There are certain survival factors that have to be taken into account when we begin to change children's behaviour in significant ways.

Perspectives on Challenging Behaviour

The **behavioural perspective** assumes that all behaviour is learned and shaped by reinforcement. Positive reinforcement increases behaviour; punishment or negative reinforcement reduces the frequency of behaviour. From the behaviourist perspective, a human being is a set of responses shaped by the external environment.

A **cognitive perspective** suggests that we behave according to the way we think, visualise or imagine. From this perspective, the human being is more than just a set of responses to stimuli; it is a conscious being, making choices, perceiving the world in certain ways and behaving according to the rules of logic laid down in their thinking brain.

The **psychodynamic perspective** conceives of behaviour as a result of unconscious conflicts, primitive drives and deep-seated anxieties or fears of which the person is consciously unaware. From this perspective, we are pawns of our unconscious minds, pushed and pulled by powerful forces beyond our awareness.

There is a new model emerging, best understood as a **biopsychosocial perspective**. This model conceives the human being as a totality of biological, psychological, and social factors, all exerting equal influences on behaviour.

We all have one of these perspectives about children's behaviour. Our perspectives become our understanding and our understanding shapes our responses. The more we become aware of our perspective, the more we can alter it; the more we can alter it the greater the opportunity to come to a new understanding and create new solutions.

The Biopsychosocial Perspective

In the biopsychosocial perspective, all behaviour is a result of brain growth and genetic, environmental, social, familial, health, parenting and hundreds, if not thousands of miscellaneous factors. All children are born with a particular temperamental constitution. This is the biological, largely genetic basis of personality. Temperament is a given and remains relatively stable throughout the life span. The behaviours we exhibit change over the course of time; the unique temperament of a person does not change much.

All these temperamental traits have value and all can be neutral, positive, or negative. Any parent with more than one child quickly notices the different temperaments of their children and gradually becomes aware how different temperaments translate into different behaviour and therefore parenting styles. Simply stated, some children are easier to rear than others and it is temperament that is responsible for this. Just as they can be either easy or difficult to rear, the differing temperamental traits of children make them more or less easy to teach. This is nature and it is this natural disposition of children that requires us to create environments at home and school that closely match temperament.

The Spirited Child

Writing in 1998, Kurcinka* describes children she refers to as 'spirited'. Using temperament as a starting point, she identifies six parameters of behaviour characteristics that make some children difficult to contend with. These are:

Intensity – powerful reactions
Persistence – not giving up easily, not changing one's mind
Sensitivity – quickly responsive
Perceptiveness – notices everything
Adaptability – uncomfortable with change
Energetic – need to be on the move

Kurcinka makes a powerful case to support the idea that it is the responsibility of the adults in a child's life to profile the child's temperament, match it to their own temperament, and create environments and interventions that facilitate a balance between the two.

Whether or not we use the psychological definition of temperament or wish to conceptualise children as 'spirited', there is growing evidence that some of what we call challenging behaviour results from biological traits and must be recognised as such. Adults must try to understand this and adapt the surroundings, expectations and teaching/parenting methodology more towards the needs of the child.

The Three Ages of the Child

Every child has three ages. The easiest one to comprehend is **chronological age** although even at this, seemingly most basic level, confusion can arise.

---

* Kurcinka, M. S. (1998). *Raising your spirited child*. New York: Harper Collins.

Every sixth class teacher knows that children at this age differ widely in physical traits and characteristics. Some are clearly well into the beginning stages of puberty; some have not reached pubescence at all. Physical differences translate into different expectations about levels of maturity and behaviour. Children who appear physically beyond their age are often perceived as being able to function at a more mature level than their brain will allow. So looks can be deceiving and it is important, as a general rule, to tailor interventions to chronological age.

The next age of the child is **intellectual age**. Intellectual age refers to the general level of intelligence of the child (IQ). Intellectual age can be greater than or less than chronological age and tends to remain stable throughout the life span unless disease, trauma or environmental toxins impacts it. Children with General Learning Disabilities all have significantly below-average IQ. This low IQ means that their level of conceptualisation, generalisation, abstraction and comprehension will be below chronological age. We don't speak to these children the same way we do to other children because of their intellectual deficits. Likewise, in the case of gifted or exceptionally able children, we match our language to their level of cognitive ability. Matching our interventions to the intellectual age of the child makes it likely we will create more effective solutions to behavioural difficulties.

The third age of children is their **emotional age**. This is where things can get a bit confusing. The emotional age of a child fluctuates with environmental factors such as stress, trauma, anxiety and health status. A child's emotional age can be well below their chronological or intellectual age. Take for example the nine-year old who throws a fit after losing a football match. He is acting like a three-year old in a tantrum.

What is important to realise is that emotional age can be below chronological or intellectual age but can never be truly above either of them. All children who appear to be mature beyond their years, who have 'old heads on young shoulders' have been socialised to act that way, and it is an act. A good example is the child from a home in which there is severe alcoholism. They are often placed in a position of caring for the parent or other sibling. Becoming 'adultified', they develop attitudes and vocabulary, a pseudo-sophistication that is deceiving. When we interact with them at this false level of development, things often go awry. As a general rule it is always advisable to intervene with a child at the level of their emotional age (remember-it can be below or equal to chronological and intellectual age, but never above it). This is especially true of discipline.

Getting the match right, perceiving the child as he or she truly is, is an important part of generating appropriate education and behaviour management strategies.

The Child's Brain

Everything happens in the brain. It is the brain that learns, the brain that remembers, and the brain that behaves. At the level of the brain there are no emotions, only chemical and electrical discharges that govern all of human life. Anything that happens to a person in life will have an impact on the brain and how it functions. A basic understanding of brain growth and development can help us understand the root causes of some challenging behaviour.

The brain grows from bottom up. The first part of the brain to grow and develop is the **brain stem**, which controls all basic life functions such as heart rate, respiration, and blood pressure. Any injury to the brain stem usually results in death. The brain stem is fully formed and developed in the womb. The next region of the brain to grow is called the **diencephalon**; developing throughout infancy into childhood, it regulates and controls motor skills, and some sensory processing. The third brain region to develop is the **limbic system**, the seat of human emotional regulation. Developing through early childhood and into adolescence, it controls, in addition to emotional functioning, memory, primary sensory integration, and attachment to significant people. The last region of the brain to grow is the **neocortex**, which does not finish maturing until early adulthood. This brain region is responsible for reasoning, problem solving, abstraction, planning and evaluation.

It is well documented that children who live in chaos, or who are victims of severe neglect, physical or sexual abuse, or who live in violent surroundings, have altered brain functions. The gender difference among these children is interesting. Boys who have suffered these events tend to be much more hyper-vigilant and active than girls, who tend to withdraw. The boys are constantly scanning the environment for any sign of threat or rejection. Once they perceive it, even if it's not real, they are prone to get angry, retaliate as if they were being threatened, run away or become abusive, perhaps aggressive. The girls, who are just as watchful as the boys, tend to retreat into silence, daydreams and an imaginary world of silence.

Alterations of brain function resulting from trauma, chaos, or neglect can cause the brain to reorganise itself and become fixed in a state of alarm and stress response. Children in our schools who live in these environments are often those who present the most challenging behaviour. An understanding of what is going on in their brain can help us recognise that they are not being wilfully difficult, not making a choice to challenge and provoke, but are subjects of brain changes they cannot control and are not responsible for creating.

The Importance of Attachment and Other Protective Factors

Attachment, the bond between the child and the significant adult, is one of the most powerful predictors of how acceptable our behaviour will be. When the attachment is secure the child feels safe, loved and loveable. Under this condition the child, particularly during the first two to three years of life, knows that adults can be trusted and will not cause harm. When attachment is insecure, the child learns the opposite; he is neither lovable nor loved, adults cannot be trusted and they will harm you, one way or the other. These inner beliefs, formed before we have language to comprehend them, shape the way we interact with the adults who are significant to us. Under secure conditions, children will want to please the significant adults in their lives and this is why they respond to warmth (or even to being scolded, when necessary). But children who are not securely attached have a badly adapted comfort zone, are not responsive to praise or scolding, and do not seem to have the critical need to please adults that is manifested by their peers.

There are **other protective factors** beside attachment, all related to brain functions, that appear important in preventing challenging behaviour. Our ability to **regulate** our behaviour, something that develops in early childhood, will protect us from losing control of our behaviour. The capacity to **affiliate** with others, to feel as though we belong to our fellow man, helps us cooperate. The ability to be **attuned** to others and to tolerate the differences among people helps us know that the world needs many excellences. Our capacity to **respect** others helps us work together in harmony in a classroom and cope with differing needs and demands. All of these functions lie deep within the brain; all are subject to interacting with a predictable environment that stimulates the brain to grow in positive ways. The absence of a predictable, secure environment leads to the reduction or absence of one or all of these functions, leading to increased likelihood of challenging behaviour.

What Are We to Do?

There are no easy, short-term solutions to challenging behaviour. Prevention is the best beginning. This begins in the early years at home, and through to infant classrooms – the focus must be on the development of an emotional vocabulary; 'feeling' words that help us talk about how we feel rather than having to act out how we feel. The revised Primary Curriculum (1999) places as much emphasis on emotional and social development as it does on academic development. We are fortunate to have ample opportunities to develop these 'protective' skills by using the creative arts curriculum, circle-time and SPHE to facilitate personal growth. For some children however, this curriculum is insufficient. They require more direct and intensive educational interventions.

Whole-school policy is the way forward. Classroom based, simplistic interventions will not solve the problem, and at their best they can only result in a temporary reduction of challenging behaviour. There are a number of strategies that can be implemented such as the 'social autopsy' (www.ldonline.org) and 'social stories' (www.thegraycentre.org) that are of proven effectiveness when applied systematically over extended periods of time as part of school policy.

We also know what doesn't work. Suspension is totally ineffective, as are punishment and sanctions. Challenging behaviour arises because children have not learned a better way to meet their needs. Reducing the unacceptable behaviour of children without increasing the acceptable behaviour will never prove successful. Sustainable, comprehensive interventions that use the curriculum, the school environment and whole school policy are the way forward.

At secondary level, the exclusion of children with challenging behaviour might generate a threat to society and should be avoided, whenever alternative solutions are possible.

The Single Most Important Person and the Single Most Important Factor

In the life of every child who exhibits challenging behaviour, there is one person that can make the essential difference and that is the teacher. A teacher who takes the time to learn about the causes of challenging behaviour, who makes a commitment to learn better strategies of intervention, to use the curriculum, adapt the classroom environment and alter methodology when necessary will prove to be the child's agent of life transformation. Nothing is more important than the teacher's ability to sustain hope that children can learn to change their behaviour for the better.

The single most important factor to instil in every child is *hope*. Hope is the internal image of a better future. An image in which there is safety, predictability, nurturance, respect, humour and flexibility in living life, helps us overcome all sorts of difficulties. Hope saves lives. There is no room in any school building for an adult who does not have a hopeful outlook for all the children they teach.

8. Adolescent Development

This article originally appeared in "The Essential Parents' Guide to the Secondary School Years", PrimaryABC, 2004 (www.primaryabc.ie)

Moody, sullen, oppositional, opinionated, lazy, sleepy; all these and more are among the common terms used to describe teenagers. Parents the world over have been struggling for the last fifty years or more to understand their teenage children. Why do they act the way they do? What happens around age twelve or thirteen to turn that cooperative, pleasant child into a stranger who doesn't seem to want to belong to the family any more? What's going on in there anyway?

An understanding of the adolescent can help us love and cherish them for all the traits that so frequently drive us mad. An adult who takes the time to learn about the tumultuous years of adolescence will soon enough recognise the factors that contributed to their own teenage challenges and will realise that most of what we perceive as troublesome in teens is in fact just the necessary developmental changes they are undergoing on their journey into young adulthood.

A Period of Rapid Change

After infancy the human being will never undergo such a period of growth and development as during adolescence. Everything is growing fast: mind, brain, body and intelligence all experience tremendous change during the teenage years. Unfortunately, unlike infancy, the people around the teenager can't see the struggle for change. We observe our infants and toddlers stumble as they learn to walk, babble as they learn to speak, wobble as they try to run, but we can't observe on the outside the similar changes a teenager is experiencing. Our inability to be witness to the hidden changes of adolescence causes us to puzzle over the external manifestations of this period of great development. When we cannot witness we become anxious and confused. It can be helpful to take the mystery out of adolescence. If we take a look inside and uncover what seems to be deeply hidden we can appreciate the intricacy of adolescent development.

We'll begin our journey with a systematic look at the major areas of change our teens are undergoing. Starting with the psychological movement of adolescence, we will proceed to look at the complexity of the human brain and it's growth during the teen years. We will then look at the influence of familial factors on adolescent growth. We will end our journey with a look at societal factors that further complicate this period of human growth, a period of such great change that it will never be equalled again during the life span.

The Psychology of Adolescence

All human life is movement through social spheres. There is no such thing as an isolated human being. Even the hermit needs a community from which to separate himself. We are born to be social beings and from the very moment of fertilisation we grow in relationship to other people. The struggle to define ourselves as social beings is characterised by our innate need to overcome difficulties and to strive always towards the positive side of life. We will come to look at the social side of the family later in this chapter but for now we must just remind ourselves that all life is movement and take this knowledge to look at the movement of adolescence.

The major task of adolescent psychological movement is to differentiate oneself from the family, to become in their own mind a unique person, not merely an appendage of father, mother, grandparents, and siblings, but a person into oneself. In order to accomplish this necessary task, all sorts of changes must take place. First of all they must have the courage to confront this task. Courage in this sense is strength of heart to begin to venture further from home, to undertake new life challenges such as work, school, sports, and social experiences. In order to manifest this courage it is essential to trust oneself. But the problem is complicated by the fact that in order to trust oneself one has to be willing to make mistakes, learn from them, and have sufficient self-esteem to cope with failure experiences that are inevitable. This requires allies, fellow travellers who they can count on to be there for them, to comfort them in their mistakes, and to encourage them to try again.

The allies the adolescent chooses in the journey to discover the unique self are peers. Who better to accompany them on this rocky road to creating a unique self than others who are travelling along the same path? The peer group becomes so overwhelmingly important during adolescence because it serves as a protecting factor against the family's need to keep the teenager in the nest. The tension between the adolescent striving for independence and the family's unconscious desire to hold on to what is most familiar leads to some interesting events. Any parent of a teenager has experienced some of the amusing, and not so amusing artefacts secondary to the movement for independence.

The family values that were once so solidly embedded in a child suddenly become signposts of rebellion. The teenager won't go to church, believes dad has "sold out" to corporate life, thinks mother is too intrusive, needs to keep secrets, etc. All sorts of contortions begin to occur. Adults puzzle over this, but once we recognise that the peer group and it's seeming magnetic hold on the teen is merely a bridge being used in the journey to creating a new self, we can confront this 'rebellion' in the light of comfort. The comfort comes from the security in knowing that in all but the most extreme cases, the teenagers turn away from family values is temporary. By the time most teens are young adults their core values have returned to those lived by the family. The rebellion of the teen, articulated by an exaggerated reliance on the peer group, is just an illusion. It is the necessary process of experimenting to discover just who you are and what you believe.

A significant part of the journey to independence is in discovering one's core values rather than merely mimicking them from our parents. Teenagers begin to think deeply about things and this thinking, as we will soon see, is not unlike the act of learning to walk. Although we can't see it, the teenage mind is stumbling from one assumption about life to another, from one moral judgement to another, and it will take a good few years for the mind to sort this out and reach some final, though in themselves fragile, conclusions. Our teenagers are beginning the process of discerning the meaning of life. In childhood the meaning comes from parents and family, in adolescence it begins to come from within. But what is "within" is being created and, as is the case with all creation, various levels of pain and struggle accompany it.

The psychological movement of adolescence is real but hidden from our view. It's helpful to keep in mind this movement and understand that most of what we see on the outside is healthy and necessary for the growth of our teenagers.

The Adolescent Brain

The neurosciences are beginning to unravel the mysteries of the human brain. The grapefruit-sized organ weighing about 2-3 pounds, having the consistency of butter at room temperature is responsible for all we undergo in life. For a long time scientists thought that by middle childhood the brain was nearly fully developed, needing only some small refinement before it reached adult maturity. Recent research is proving this wrong and leading to some interesting discoveries about the adolescent brain.

Our teenagers are a work in progress in more ways than we can see from the outside. The physical changes of puberty take place in the span of a few years but the changes in brain structure, chemistry, and biology take many more years to solidify. That gangly adolescent is even ganglier on the inside, particularly inside the brain.

What part of our brain makes us uniquely human? It's the part on the outer layer, in the frontal lobes of the brain that connect with the inner parts and permit us to manifest what is truly unique about our species. The frontal lobe of the human brain mediates complicated functions such as delay of gratification, ability to tolerate frustration, planning and evaluation of plans, thinking about emotions, controlling of impulses, and learning about life by generalising and drawing conclusions from our mistakes. It is the role of this part of the brain to connect with the deep brain centres, where emotions are generated. In these centres deep inside the brain our emotions are like wild stallions. It is the function of the frontal lobes to put the bridle of reason on emotion.

What relevance does this have to our understanding of the teenager? It has tremendous relevance because we now know that the fibres that connect the frontal lobes to the emotional centres within the brain are not fully mature in early adolescence. In fact, there is mounting evidence that they don't mature fully until early adulthood. If these fibres are not mature, then the mediating functions of the frontal lobes are going to be inefficient in controlling emotion and in making plans, evaluating these plans, and revising them in the light of new data. Although all of the above is oversimplified it does take some of the mystery out of adolescent behaviour. For example, the struggle of so many teens to learn to plan ahead in completing school projects, to understand how to break plans down into manageable units, to revise plans in the light of impending deadlines, all can be seen from the viewpoint of brain development. What may seem like laziness, lack of motivation, and the opposition to study, suddenly becomes comprehensible. There are more teens that have brains undergoing change that will take years to mature than there are 'lazy' kids who don't know the value of education.

Society and the Teenager
Adolescence is a rather recent 'invention' for human beings. One simply went from childhood to adulthood. In fact, in many indigenous societies there is no period of adolescence. As the child grows strong and is able to contribute to the community's survival, he or she assumes adult responsibilities. As society has become more complex and legal systems evolved to prohibit child labour, a new view of childhood emerged. This view created a period of preparation for adult life, a period in which children would serve a sort of 'social apprenticeship'. This has evolved into what we now call adolescence. But as society has become more complex, so this process of apprenticeship has become more complex too.

Most adults today have lived in a world of moral surety. There were figures in society invested with tremendous authority. The church, the legal system and our parents and teachers all held moral and physical superiority over us. They told us what to think, what to do, where and when to do it. The consequences of violating these imperatives were swift and sometimes severe. Be it corporal punishment, moral damnation or removal from school, everything in society reinforced the inferior role of the teenager.

No one needs to be reminded that this has all changed. The church has lost its position of authority, school teachers have fallen off their pedestals, parents are no longer dictators in the home, and the result is that the teenager lives today in a world of shades of grey where once they lived in black and white.

Imagine what it is like for our teens today. Undergoing the natural movement towards separation from parents towards individuality, having a brain that is slowly developing, they are living in a world of moral ambiguity with which they are poorly equipped to cope. Even society's over-arching authority, government, is no longer perceived as a source of goodness. Confronted on every side by shades of grey, our teenagers are increasingly at the mercy of marketing hucksters telling them that significance and individuality come from purchasing the right clothes or most modern mobile phone, and from having a car at age 18!

What is "Normal" Teenage Development?

Adolescence is a period of struggle and growth, most of it happening on the inside. What is visible on the outside is a symptom of an unfolding process of change. It is normal for teenagers to be moody. Most of the mood swings of adolescence are a result of rapid metabolic changes, not emotional turmoil. It is normal for teenagers to be oppositional. What appears as opposition is a testing of the waters of parental love (Do they love me enough to let me be a person on my own rather than an appendage of my parents?). It is normal for adolescents to take extreme moral stances and see the world as morally black and white. This is especially true of the younger adolescent – the brain develops in such a way that subtle moral thinking doesn't usually occur until late adolescence. It is normal for teenagers to become 'infatuated' with a love object. Adolescent love is rehearsal for mature adult love (the bridle of reason must be put on it however). It is normal for the adolescent to turn away from core family values. By young adulthood the family values are usually back in place. It is normal to test adult authority. This is the teenagers attempt to discover how ready they are to face the world on their own.

A lot of what we expect, and what we do to our teenagers is inappropriate. We rush them to make decisions their brains cannot make at too young an age. We expect them to function at levels of maturity that their brains are not able to reach. We use the same methods of discipline at age seventeen we used at age seven, expecting blind compliance from a person engaged in the struggle to become an individual capable of making their own decisions.

Getting it right needn't be so difficult. With a basic understanding of adolescent development we can change the way we think about our teens. When we have changed the way we think, we can accept them for what they are – a work in progress that needs a good bit of trial and error in order to progress. I think it might just help adults if they realise that the toddler who stumbled and fell while learning to walk is hidden inside the adolescent. Stubborn, afraid of being seen as inept, they make mistakes in living and learn from them. Our job as parents of teenagers is to encourage the making of 'better mistakes' and to stand by them with understanding of their development and a deep love of the journey we are on together.

Special Education and the Secondary School Student
This article originally appeared in "The Essential Parents' Guide to the Secondary School Years", PrimaryABC, 2004 (www.primaryabc.ie)

9. Hidden Disabilities that may emerge in Secondary School
This article originally appeared in "The Essential Parents' Guide to the Secondary School Years", PrimaryABC, 2004 (www.primaryabc.ie)

Statistics indicate that between 7% and 10% of all students in secondary school have a special education need (SEN). The incidence of these needs varies from community to community and is susceptible to variance due to definition, availability of assessment, and terminology. One thing is certain however, a significant number of students in secondary school have special education needs. Educating secondary school students who have a SEN is complicated by a number of factors:

- The age of the child may militate against successful remediation or treatment
- The complexity of the curriculum and the demands of the examination system make it difficult to get academic results of a high quality
- The limited availability of remedial support reduces the opportunities for assistance
- The tendency of some secondary school teachers to be inflexible and unwilling to differentiate their teaching style places the student with a SEN at a disadvantage for learning

The subject of special education is so wide as to be almost impossible to encapsulate in a short chapter. Because of this the focus of this article will be on the **'hidden disabilities'** that seem to arise when a student moves from primary to secondary school. These hidden disabilities can be a puzzle to parents and teachers alike. They can be just as puzzling to the student who previously had only minor school difficulties but who now finds it increasingly difficult to comprehend material and master content. Although they are hidden they are real, and real special education needs require real solutions.

What Are Hidden Disabilities?

Primary school is a place where children are provided structure and support; the learning experiences are repeated in a sequence of increasingly complex demands and thoughtful guidance is provided by caring teachers and teaching assistants. Children who struggle and have academic or behavioural difficulties in primary school are usually spotted quickly by their teachers, and reasonably good support systems are currently in place in Irish primary schools. Classroom teachers have been taught how to change their teaching style, methodology and resources (especially the younger generation of teachers). All of this makes it likely that those children with special education needs are recognised early in their primary school years and that appropriate supports are provided.

Part Three

Despite this there exists a form of special education needs that seem to lie dormant throughout the primary school years, only to manifest themselves during secondary school. Children who appear to be fluent readers suddenly have difficulty comprehending textbook passages. Students who never struggled at mathematics experience problems with calculus, geometry and trigonometry. Attention and concentration, which were once never an issue, suddenly are notably lacking. Memory that was once intact now seems to be faulty. What is happening to change students who were once performing adequately into students who struggle to learn?

The human brain is an organ that grows and develops at its own pace and in a structured, patterned and predictable sequence. The brain is also an organ that is constantly interacting with the external environment. The demands placed upon the brain must match its developmental capacity to meet those demands. Once the demands of the external world exceed the brain's capacity to cope, trouble begins. This is an over-simplified explanation of the term 'hidden disability' but a useful one just the same. For most children the progressively slow pace of the primary curriculum exactly meets their brains ability to cope. The pace of primary school lessons, the number of years being taken to complete primary school, the care and structure provided by primary school teachers all combine to provide an appropriate external environment for the child's brain. Most cope well within this structure.

There are some children, however, whose brains are vulnerable, but not vulnerable enough to bring them to the attention of primary school teachers. The reason for this is the relatively good match between the curricular demands in primary school and the capacity of the child's brain to meet those demands. All things being equal, the curriculum gradually becomes more complex as the brain becomes more capable of handling the increased complexity.

Pace of change and its impact on students

Things can change quite rapidly once the transition to secondary school is made. First of all the pace of lessons accelerates tremendously. The need to prepare for state examinations and the shortness of the school year means that teachers have to compress a huge amount of information into a small number of learning opportunities. The level of comprehension required to read textbooks becomes greater with each passing year of secondary school. The number of textbooks that must be comprehended increases exponentially as well. The demands of the curriculum take an exacting toll on mental capabilities. The number of teachers and the number of teaching styles vastly exceeds whatever was experienced before. It is no wonder, given all the above, that students who never had learning difficulties in the past seem to appear out of nowhere as needing more assistance than any classroom teacher can provide.

This interface between the brain's capacity to learn at different ages and the incredibly complex nature of the secondary school curriculum can overtax the vulnerable brain. When that happens subtle and sometimes not so subtle, learning difficulties become apparent. Though these vulnerabilities were present all along, it was the lack of critical demand placed on the brain that caused them to be hidden. The so-called hidden disabilities are real and require intervention. Unfortunately, since they seem to occur out of nowhere parents and teachers often perceive them as laziness, willful malingering, lack of motivation, or stubbornness. This misinterpretation of real disabilities often results in blaming the student instead of providing support.

THE DIFFERENT TYPES OF HIDDEN DISABILITIES

- **Reading Difficulties**

   Many students have difficulty with reading and not all reading difficulties are secondary to hidden disabilities. How can you tell the difference? Well, the surest sign that there is a hidden disability is when it seems to arise out of nowhere, when you haven't been informed during the primary school years that your child may have, or has, a reading difficulty. There might also have been some early indicators of potential reading disabilities such as significant left-right confusion, delayed speech, fine motor or gross motor control problems*, reluctance to draw or colour. But the acid test of a hidden reading disability is its sudden appearance in secondary school. A student may complain of finding the books 'boring', of having difficulty remembering what they have read, or just show signs of being reluctant to read at all when they previously were interested in reading books at home.

• **Mathematical Difficulties**

The sequential nature of the maths curriculum in primary school, combined with the new emphasis on active learning experiences, makes it difficult to discern which students may have a hidden maths learning disability from those who are reluctant at maths or who have a history of poor teaching of mathematics. In secondary school the maths curriculum becomes increasingly abstract. It escalates in complexity at a rapid pace and requires considerable mastery of what was learned before in order to succeed at higher levels of complexity. Students with a history of spatial-perceptual processing (difficulty organising themselves in time and space), may have an underlying processing deficit in the brain that will make higher mathematics problematic. Any sudden, unexplained deficit in maths learning could also be a result of poor teaching so judgement must be suspended until some assessment has been completed.

• **Attention and Concentration Difficulties**

In secondary school, lessons and the methodology of teaching tend to become centred on lecturing and verbal input. Children who were used to the more experiential mode of learning characterised in the Revised Primary Curriculum have not been accustomed to sitting for long periods of time and listening (at least they aren't supposed to be accustomed to this). As a result, the demands placed upon the 'attentional' systems of the brain were never stressed beyond their

---

* Fine motor control is the control of small precise movement (pointing to a small item with one finger). Gross motor control is the ability to make large general movements (waving an arm).

functional capacity. Suddenly, and in the space of only two months, the brain has to absorb 40 minutes of verbal input, place it into storage, recall it later, sort it for its essential components, and place it into long-term memory. Students with attentional deficits often come to the attention of teachers and parents during secondary school.

- **Social Difficulties**
  Going through primary school from year to year with the same group of children helps pupils feel secure. They are able to anticipate all possible permutations of behaviour from classmates and know to understand teachers' level of tolerance. Upon entering secondary school everything changes. A new peer group presents new challenges to social affiliation. Up to eight or nine different teachers a day, each with different expectations and different teaching methodologies must be encountered. Although we are programmed by evolution to be social creatures, nothing in our young lives prepares us for the complexity of social interactions we will meet on a daily basis in secondary school.

- **Emotional and Behavioural Difficulties**
  Unfortunately there are some forms of significant behavioural and emotional disturbances that do not appear until adolescence. There is a form of conduct disturbance, one of the most severe of childhood behavioural disorders, that does not occur until mid-to-late adolescence. The child who previously was cooperative and pleasant seems to turn overnight into an incredibly provocative, rule-violating individual. Some psychiatric disorders do not manifest themselves until adolescence and among the most severe is schizophrenia. Deviant social behaviour under the influence of alcohol or drugs is also a possibility.

What To Do If a Hidden Disability is Suspected

The fatal mistake is to ignore the problem. Any *sudden* change in learning style or proficiency, or a sudden change in behaviour or socialisation is a possible warning sign of a hidden disability. There are, of course, a host of factors that could be causal, including family difficulties, adjustment problems to secondary, a change of housing during adolescence, and death or loss of a loved one. A reasonable guideline is that if a change is noticed and lasts for over three months, it needs to be investigated. Hoping the problem will go away by itself will only lead to further complications and more time needed to treat the problem.

Assessment is the key to discovering whether or not a student has a hidden disability. Assessment must be completed by a professional who has the clinical knowledge to know what to look for and where to look for it. Assessment can be completed by a variety of professionals including qualified special educators, psychologists, speech and language therapists, psychiatrists, paediatricians, (depending on the age of the student) or neurologists. But assessment is only the first step in the process of correcting the problem.

Assessment is only useful if it leads to corrective planning. A diagnosis alone will not deal with a hidden disability. Assessment is a tool to understanding the problem and creating a treatment plan. The plan must be implemented by competent professionals in an educational or clinical setting. It is imperative that all work together in the interest of the student. Isolated interventions are hardly ever successful. Parents must ensure that any assessment will be shared with educators with a view to creating an effective treatment/intervention plan. It is also imperative that parents and educators work together in a spirit of open communication and collaborative planning.

What to Look for in a School

If a student has a hidden disability and assessment has indicated a need for intervention, there are some critical issues that must be confronted in order to proceed. First, the available resources in the school must be scrutinised.

- Does the school have a Learning Support Teacher or a Resource Teacher on the staff?
- If so, how many students are assigned to each caseload?
- What are the maximum number of hours a specialist teacher can provide the student?

- Does the teacher have a higher qualification in special education?
- How willing are subject area teachers to modify their methodology or resources?
- Can structured support for homework be provided?
- What are the attitudes of the teaching staff to the student's difficulties; do they view him or her as lazy and unmotivated as opposed to being a student with a learning difficulty?

What to Look for In Your Family

- Do you, as parents, have an accepting attitude to your child or do you perceive them as lazy and unmotivated?
- Is there an opportunity for you to discuss the problem with the child and come to consensus about the nature of the difficulty?
- Will siblings be enabled to understand that a genuine problem is present?
- Will the extended family come to the same understanding?
- Can you work cooperatively with teachers and other professionals?
- Are financial resources available to provide private tuition or therapy if necessary?
- Can you, as parents, accept your child as they are without criticism, blame or guilt?

Conclusion

So-called hidden disabilities are real disabilities. Although they sometimes seem to arise out of nowhere, there is often a history of developmental delays and early warning signs. Hidden disabilities can be overcome with proper assessment, treatment and intervention. This is more likely to be successful when all concerned work together cooperatively. An educational environment that is accepting of differing learning styles and learning needs provides the best opportunity for overcoming a hidden disability.

10. The Underachieving Adolescent

This article originally appeared in "The Essential Parents' Guide to the Secondary School Years", PrimaryABC, 2004 (www.primaryabc.ie)

Why is it that some teenagers perform in school, and sometimes elsewhere in their lives, well below their level of capability? What happens to children when they leave primary school, make the transition to secondary school, and suddenly seem to lose all interest in learning, studying, preparing for exams, and complain that school is "boring"? For many parents the underachieving adolescent is a puzzle and a disappointment at the same time. This article takes a look at the factors that lead to underachievement and offers some suggestions about what can be done about it.

Pathways to Failure

The transition from primary to secondary can bring several issues. Primary school tends to be a place where children are structured, nurtured, and looked after. However, secondary school can be a harsh and unforgiving place where an impersonal atmosphere prevails and where high expectations to "get with the programme and plan ahead" are the norm. Many children enter secondary school with a history of poor study habits, poor preparation, and poor planning that primary school teachers have accommodated to some degree. These habits are suddenly no longer tolerated upon entry to secondary school. The pupil, who is not supposed to be a student, is left to his own devices to try and succeed. A significant number of them lack the inner resources to do so.

At the same time the adolescent is struggling to cope with the greater demands of secondary school, he is confronted with the full force of developmental changes of the teenage years, discussed in a previous chapter, see page 80. Needing to make educational and career decisions at a young age, an age at which the brain is often unable to cope with these demands, the teenager finds himself caught in a cycle of difficult external demands and undeveloped internal resources. This is a difficult enough process for the average teen but for the underachiever it can become an insurmountable series of obstacles.

The combination of the pressures of adolescent development and the demands of secondary school often produce the classic signs of **underachievement** outlined in Rathvon's book, *The Unmotivated Child*, and which include the following:

- Failure to develop meaningful educational and career goals
- Engaging in a pattern of self-defeating behaviour
- Withdrawing from family and friends
- Attachment to an underachieving peer group
- Withdrawal from competitive and evaluative situations
- Feelings of anxiety and depression about the future
- Turing to external sources of gratification (drinking, drugs)

The demands of secondary school can be as tough on the family as they are on the teenager. Faced with the well-founded concerns, and sometimes complaints, from teachers and year-heads, parents become increasingly frustrated with their teenage underachiever, family rows develop, tensions increase in the home, and what was once a loving family becomes a battleground. What is happening inside the teenager to create all these difficulties?

A Look Inside the Mind of the Underachiever

All of us have thoughts, beliefs, and attitudes that were constructed early in life, usually before we had language to describe them to ourselves and to others. These inner "maps" of the outer world consist of images of others and ourselves that are fairly accurate reflections of what we see in the external environment. This inner world of maps and templates are called "working models" of the external world. Most of us have flexible working models and can adjust them as new situations arise. However, for the underachiever the inner world can be a place of rigid demands and badly adapted beliefs. These can persist even when positive experiences are occurring. It is well known that children reared in stressful environments who have been subjected to harsh corporal punishment, psychological abuse (being told they are stupid, no good, not good enough) often grow up to become underachieving adolescents and adults.

There are many factors that contribute to this type of badly adapted 'working model' of the world and teenagers tend to speak in a sort of coded language. Statements from underachieving children such as, "School is boring", "I'm no good at maths", "The teachers don't care", are all ways of attempting to explain what lies hidden: their working model of themselves and the world. They can however be understood by parents and teachers once we learn how to translate this language into ordinary English.

"It doesn't matter if I study or not. I studied for that exam but failed anyway. I told you, it just doesn't matter!" This statement, heard by many a parent and teacher, is the adolescent's way of telling us they feel as though they have no control over their lives. This feeling of no control is often associated with the state of being highly disorganised. The desks, room or lockers of these teens are more chaotic than those of their peers.

The belief that they have no control of what happens leads them to stop trying to control what happens. The working model of these children prevents them from asking the one thing they most need to ask: "Please help me learn how to control my life, organise myself internally and externally, and to become a successful student."

Teenagers who are chronic underachievers are prone to create catastrophes where none exist. They anticipate failure with exquisite success. They are unable to make the link between preparation and success and therefore believe that nothing matters at all, everything will lead to failure no matter what they do. Having so few inner resources and being dependent on the developmental limitations of the teenage brain, the underachieving adolescent walks through life from one failure to another. These seemingly endless failure experiences have a disastrous effect on self-esteem. Believing that nothing they do leads to success they complicate the matter by *believing* they are failures at school and home. Because our success experiences feed self-esteem, teenagers who have frequent failure experiences learn they are just not up to the task of life, become depressed and feel hopeless. They get bogged down in a cycle of anxiety, failure, damaged self-esteem, more anxiety and more failure. It is a rut that if not interrupted can have disastrous life-long effects.

Turning away help

The underachieving adolescent is caught in another trap: the need for help and assistance and the turning away from it when it is offered. This is an inner dilemma which can be characterised by constant rejection of parent or teacher offers of help. Feeling helpless, inadequate, and lost in a world they cannot comprehend, the underachieving adolescent both wishes for and fears assistance. Indeed, these teenagers can perceive assistance, when it is offered, as some sort of "trick" perpetrated by adults. The underachieving teen has often been reared in a harsh and unforgiving family environment where a belief about innate worthlessness has been unconsciously put into their mind. Given the core belief of being both unloved and unlovable, these teens will quickly perceive any offer of assistance as just another ruse by adults who, they believe, will abandon them at the next failure experience. Why accept help if the helper is only going to insist on instant success?

Unfortunately, the underachieving teenager has such well-entrenched internal beliefs that even success is interpreted as failure. Many teachers and parents have puzzled at the teen that states, after having succeeded in a task at school, "I got lucky, they asked me the things I knew." This seemingly paradoxical statement must be interpreted for its real meaning, "I'm stupid and unable, and if I succeed it is a matter of luck over which I have no control." The combination of low self-esteem, expectation of failure, the need for help complicated with the rejection of offers of help, all create an exceptionally complex inner working model of a harsh and unforgiving world. The difficulty of parents and teachers to understand this inner world, the harsh and inflexible demands of secondary school and the normal developmental curve of the adolescent brain, all conspire to keep the underachieving teen down at the bottom of the pile.

Part Three

What happens to the underachieving teenager? Some of them leave school early after a seemingly endless series of failure experiences. Some are encouraged to leave school early by parents and teachers who unwittingly (and sometimes wittingly) send them messages that they don't have what it takes to be an academic success. Some of them, especially those whose working models of the world make them provocative and challenging to the adults around them, are put out of school. Many are just assigned to the scrap heap of teenagers about whom adults have exceptionally low expectations.

The provocative underachiever

A word needs to be said about the provocative and challenging underachiever. Although only a few fit this profile, it is probably the least understood by parents and teachers. What makes a teenager who was once a complaint or cooperative child, become so antagonistic in secondary school? In order to answer this question we must remind ourselves about the working model these teens bring to the external world. The belief that one is unable for the tasks of life, that one is both unloved and loveable, that one can never be a success and that success, when it occurs is a matter of sheer luck, is a combination leading to unpleasant results. These teenagers go through life with a sense of worthlessness that leads them to seek external proof of their internal beliefs. "I am no good and I will prove it to you" is one of the core beliefs of the provocative underachiever. "All adults are unforgiving, pain-inflicting people who will abandon or hurt me, and you are no exception", sums up the inner belief of the provocative adolescent.

These are the teens that "bite the hands that feeds them" and turn against the adults who are trying to help. None of this behaviour is deliberate and wilful. It all stems from the inner world these teens bring to every one and every thing around them.

Helping the Underachieving Adolescent

Becoming an underachieving teenager took a good few years and didn't happen in a week; therefore change will be slow, perhaps laboured and characterised by a bit of back-sliding ('one step forward, two steps back'), **but change and progress can occur**. The goal of change is a "…permanent, positive modification of maladaptive ways of thinking and behaving…". How long it takes depends in large degree to the adults capacity to support themselves and the teenager as they strive to improve things. Communication between home and school is essential and even more essential is the quality of that communication. Complaints about the underachieving teenager from either side are not helpful or productive. In fact complaints only make matters worse and prove to the underachiever that their working model is in fact correct.

Parents and teachers often mistakenly believe that better grades and academic performance are the first signs of success. In reality social behaviour almost always improves before academic behaviour. This includes a drop in the number of arguments with parents and teachers, a more cheerful and positive attitude, a higher level of energy, increased interest in activities that were abandoned previously, and an acceptance of adult feedback, supervision, and positive criticism when it occurs.

What can we, as parents and teachers, do to help?

First of all we must recognise that backsliding is inevitable. Along with this recognition comes the action of never withdrawing support after the first success experience(s). Early withdrawal of support and assistance is a terminal mistake. Remember, no matter what the source of success, the underachiever believes they are a failure. Respond constructively to backsliding; don't panic. Focus on the effort the underachiever is making rather than the product of that effort. Avoid giving bribes and rewards for success. It's the effort that counts and we need to recognise that sustaining effort is difficult for the best of us, incredibly taxing for the underachiever. Have an unshakable belief in the underachiever's ability to overcome their difficulties. Expect improvement over time, not instant results.

Teachers and the underachiever

Teachers have their own unique difficulties with the underachiever. Far too often they take student success as a sign of their own professional expertise and success. In addition, the crushing weight of the curriculum, the demands to generate good examination results, and the pressurised atmosphere of the secondary school, all combine to make it difficult for even the most understanding teacher to be sensitive to the needs of the underachieving student. Waiting for so long for the underachiever to improve can be a frustrating experience. This frustration sometimes leads to the teacher greeting the first signs of change with exasperation or disbelief that it has taken so long to occur. This can undermine the student's commitment to change. The teacher who states, "Will wonders never cease", "I wish you did this a long time ago", or "What took you so long", only make it more difficult for the underachiever to improve. It is especially damaging if such statements are made in front of peers.

Teachers are incredibly influential in the process of changing an underachiever's pattern of behaviour. Like the parents it is important to focus on the effort to change rather than on the results of this efforts. Taking the time to offer encouragement at even small steps towards improvement is vital. Although it may seem unimportant, and may go unacknowledged by the student, being positive on attitude change or behaviour change is critically important to the student. These teenagers expect adults to be unsupportive and are especially susceptible to even small signs of disapproval. Working cooperatively with parents is also helpful. Taking the time to communicate to the parents when a teacher has observed a change for the better can yield long-term results. Letting parents know when progress is observed, or better still, when effort is being made, is especially helpful. Parents and teachers working together for the good of the underachieving student is the best method of sustaining progress.

Conclusion

The inner world of the underachieving adolescent is a world of low expectations of self and others, of diminished self-esteem, of fear of failure and fear of success, and of a belief that he/she is unlovable and unloved. It is also a world where adults are expected to be unforgiving and pain-inflicting people who say one thing and do another; who deceive you into trusting them only to abandon you when you fail. It is a world of self-fulfilling prophecies that can doom one to a lifetime of struggle, anxiety, and failure. The way out of this maze is not easy or quick. The way out can only be found with the assistance of parents and teachers who refuse to label a person as lazy, stupid, disrespectful, unmotivated, or malingering. For these reasons the lifestyle of the underachieving adolescent can be filled with pain and depression. The stakes are high, the outcome often bleak.

There is no room for loss of hope or removal from school for these teenagers. What is needed is understanding, perseverance and expectations of eventual success. Improvement takes time, sometimes a lot of time, but improvement is a reasonable expectation for the underachieving adolescent.

11. Partnership in Special Education

Parent-teacher meetings are often stressful encounters. The aims and objectives of parenting sometimes seem as foreign to teachers as the aims and objective of teaching sometimes seem to parents. There are additional tensions on both sides of the desk. The parent was once a student in school, perhaps even the same school. Parents bring old anxieties, fears, joys and disappointments into the school. They carry with them ideas and attitudes about what 'good teaching' is and how 'good teachers' act. On the other side of the desk sits the teacher, who himself or herself may be a parent. The teacher carries within herself the same feelings of having children and worrying about their educational progress, the same memories of being in school, and the same notions about what constitutes 'good teaching'. When these powerful forces meet, good communication can be difficult. It may be productive to take a look at what parents want from teachers and what teachers want from parents.

What parents want from teachers?

All parents have the same expectations, hopes, and dreams for their children. This is true regardless of social-economic status, ethnicity, religion or life circumstances. Parents want children to be confident and happy, to learn to read, to calculate, to spell, to do more this year than they were able to do last year and to grow up to live a better life than they themselves are living. These simple hopes on the part of parents then translate themselves into some unspoken expectations that are projected onto teachers and school principals. McEwan (2005)* outlines what all parents want from teachers:

- Instructional leadership
- Effective teachers
- Student achievement
- Communication
- Safety and discipline
- Involvement

Instructional Leadership

The school principal is the instructional leader of the entire educational team. Parents want school principals who:

- Have educational vision
- Are thoroughly versed in the curriculum
- Can talk in plain English, not jargon
- Drives children's educational achievement to higher and higher levels
- Can motivate all the teachers to share a common vision to strive for educational excellence at all times and who never give up on a child.
- Listens to their concerns about their child
- Keeps abreast of modern trends in education and international best practice in special education, and has a life-long pursuit of knowledge and learning

Effective Teachers for Their Children

More than anything else, parents want effective teachers. At the most simplistic level they want a teacher whose absenteeism isn't interfering with children's progress; a teacher who is on time each morning and who

---

* McEwan, E. K. (2005). How to deal with parents who are angry, troubled, afraid, or just plain crazy. Thousand Oaks, CA. Corwin Press.

doesn't rush out of the building when the children are leaving (or before they leave!). Parents want teachers who are respectful of the children and who demand respect in return. They appreciate a teacher with a sense of humour who knows when to let little mistakes go unnoticed and minor behaviour problems go unpunished.

Parents want to be told their children are wanted and appreciated in the classroom. Parents want to see notes and letters sent home from teachers that are well written, well organised and do not appear as though they were written in haste. Parents appreciate teachers who take the time to engage in continuing professional development throughout their career. Parents especially want teachers who are willing to learn about special education conditions and who are humble enough to realise that sometimes parents know more about the child's conditions then they themselves might.

Parents are especially appreciative and respective of teachers who demonstrate character through involvement in school activities, community activities, sporting events, cultural events, etc. In other words parents want a teacher who is a well-rounded, sound individual who commands respect while giving respect.

Student Achievement
All parents want their children to achieve in school. Parents want children to have meaningful homework that is a review of something important or a consolidation activity for a skill that must be mastered. Parents want their children to have a fair chance to learn, to be called on regularly in class, to be allowed to participate in the entire life of the school. Parents tend to focus on reading and maths, spelling and Irish but they also want their children to learn about history, science, the environment and the arts. Parents want children who will leave school as well-rounded people. Parents know that achievement in school has a direct relationship to occupational opportunities in Ireland. For this reason parents expect the teachers to provide everything that is essential to being able to become a productive member of Irish society. Parents also want children well

prepared to do their best Leaving Certificate exam so they will have a chance to further their education if they wish.

Communication

Parents crave meaningful communication. They are not satisfied with the occasional school newsletter (if there is one). Parents want to know what is happening in the academic life of the school and their child's classroom. They want to know about innovative projects, activities, challenges and opportunities that involve their children. Many parents want the specifics; want to know exactly what is being mastered this month or term. Parents also want to know how they can get involved in the education of their children. They want real involvement – not just fund-raising or campaigning for a new building – in practical ways that can help their children and their children's school become a centre of innovative learning.

Safety and Discipline

Parents want appropriate safety and discipline in school. They want school environments that are emotionally safe as well as physically safe. This means they do not want their children humiliated by other children or by a teacher, called names, made to feel inept or foolish, or harshly punished in front of other children. Parents want a school that is welcoming, warm, inviting, and secure for the entire school community, parents, teachers, and children alike. Parents look for teachers and principals who are not afraid to confront difficult problems of behaviour with innovative solutions rather than quick fixes (such as suspension or efforts to expel). No parent will trust a teacher or a principal unless all the above criteria are in place.

Involvement

Involvement is the two-way process of actively improving the school climate and atmosphere and academic outcomes. Genuine involvement is real partnership working towards mutually agreed goals for children. Parents want a school that is a community centre of learning and activity throughout the year. Parents are willing to get involved to extraordinary levels when they see that their involvement is appreciated, that their input is valued and thoughtfully considered, and that their suggestions are implemented in a spirit of 'let's try it and see how it works'.

What do parents of children with special needs want?

Parents of children with special needs want all of the above and more. They, like other parents, might not be able to articulate their needs clearly, they may not even be fully aware of their needs, but there are things they want just the same. Parents of children with special needs want teachers who have taken the time to learn about their child's condition, its signs and symptoms, its educational implications, its behavioural implications and its impact on learning and development. These parents want teachers who are committed; body and soul, to the education of all children, regardless of level of ability, special need condition or physical appearance.

They expect teachers and principals to be willing to learn from them, the child's parents, and to benefit from what they have learned by experience; to benefit from what the parent's know works best and what doesn't work at all. They expect teachers to be willing to adapt the curriculum, their teaching strategies, the materials and resources for a child with special needs. No parent of a child with special needs will ever respect or trust a teacher who refuses to try new things, who refuses to communicate regularly, or who insists on teaching the child like all other children.

What do teachers want from parents?

Although parents perceive teachers as self-sufficient professionals possessing a body of knowledge that parents don't have, they often overestimate the confidence a teacher possesses. Parents often forget that teachers are people too; they have joys and sorrows, successes and disappointments in and out of the classroom and are subject to the same insecurities that plague all people. Most parents fail to recognise that a teacher comes to school day in and day out, regardless of the family difficulties left behind that morning, cares for up to thirty children all day long and then has to return to the family home, listen to spouse and children, pay the bills, and start all over again in the morning. Teachers

Part Three

have needs and some of those needs they want met from parents. Teachers want the following, though the list is no more exhaustive than that which related to parents:

- Knowledge
- Communication
- Allies
- Structure
- Acceptance and support
- Respect

Knowledge

Teachers know that no one is more aware of the child's personality, learning style, behavioural profile and history than the parent. All teachers want to know about the children they teach. They want to know their interests and hobbies, their favourite toys, who lives at home with them, with whom they get along with and why and anything else about the child that will help the teacher in his or her work on a day-to-day basis.

They want to know the plusses and minuses about the children they teach. They want to know what is going well and not so well as far as schoolwork is concerned. Teachers also want to know about family matters that may affect a child's learning profile or behaviour in school. Good teachers keep this information as a source of knowledge about how to let little things go, how to draw a line in the sand when necessary and they do not use family information as a source of staff room gossip. Teachers are in the knowledge industry and it is knowledge about the child that comes from the parent that will be of the utmost help to them.

Communication

Knowledge about children can only be derived when there is open and respectful communication between parents and teachers. Communication must be regular. If a child is having success at home in overcoming some obstacle or in learning new skills, the teacher needs to know as soon as possible. Likewise, if a child is having difficulty the teacher needs to know. In order for effective communication to exist, both sides must listen to one another. For parents of children with special needs, teachers and parents need to discuss what routines are effective, what sort of tone of voice to use under which circumstances, when to be firm and when to be flexible and permissive. Teachers respect parents who communicate openly but do not burden the teacher with problems over which the teacher has no control. Teachers don't want to be therapists and psychologists but they do care about children and value open communication when it is in the best educational interest of the child.

Allies
Nothing is more upsetting to a teacher then when a parent undermines their authority and significance, either inadvertently or on purpose. After all, the teacher is the parent throughout the school day to up to thirty children (sometimes more) and must exercise reasonable care while constantly thinking about what a 'reasonable parent' would do under the same circumstances. Teachers want parents to back them up, not blindly but thoughtfully and with respect to the demands of the job to be done in the classroom. Teachers want parents to tell children that it is the teacher's duty to scold at times and to reprimand when necessary. Teachers want parents to tell children that there are times when teacher must be obeyed, even though it may seem unfair. Teachers want parents to tell children that they can talk to teacher about matters affecting learning and discuss issues of right and wrong in the classroom. Teachers are especially pleased when parents raise children who will inform them if things are going wrong in the classroom, such as bullying or intimidation. An ally is not a friend but rather is someone with whom you can work in trust and good faith, secure in the knowledge that doing what is right is always the highest goal.

Structure
All children crave structure. For children with special education needs nothing is more important than structure. In order for a teacher to provide appropriate structure for children, they must learn from the parents about the sort of structure that exists in the home. Children with autism or with ADHD require routines that are predictable in order to free them from crippling anxiety. It is the parents who provide this structure

and who must inform the teacher what systems are in place to assure that it is provided. Additionally, schools have structures to assure that education proceeds as smoothly as possible. Teachers need parents who will respect the schools' structures.

Children thrive under structured conditions at home and in school. Teachers and parents who work together to respect family and school structures and modify them to support the child's learning, are creating the conditions for life-long success in home, workplace and community life.

PART FOUR

Understanding the context, Using the Curriculum, and Interventions

This section is 'need to know' for teachers, 'nice to know' for parents.

Understanding the Context of Special Education

There is no *single* cause of special education conditions. The many different categories of special need all have different causal agents – some have no known causal agents whatsoever! Some conditions are relatively easy to understand from a medical perspective rather than an educational perspective; others are easy to understand from an educational perspective and not a medical perspective. As long as there have been schools, there have been children who do not seem to learn the way other children learn. Teachers, parents and others have always been trying to understand the nature of this difference.

Physical illness, trauma, chromosomal or genetic conditions, environmental toxins or any neurological dysfunction can all result in a special education condition. At the same time, some conditions such as the specific learning disabilities, have no known cause yet. We are a long way off from arriving at definitive answers to the question of causation in all cases.

Unfortunately, far too often this pathway to understanding leads to a requirement to **label** and **name**. Often too, the child is 'blamed' for having a 'problem'. In addition, educational systems are not particularly good at looking at *themselves* as being the causal agent of some learning difficulties.

While attempting to discover the causes, labels are being generated to 'describe' the condition. But labels are not benign – labels have an impact on people. Despite knowing this, we still feel compelled to label children. Why is this? Perhaps it's because no one likes the unexpected and the unpredictable. This is particularly true of people responsible for the education of children. When an educator can't predict what a child will

Part Four

do, or how a child will learn, he or she begins to get anxious. As a way to control this anxiety, we come to use various labels to describe the child.

No matter what the label is, the end result is the same. The educator (and parents) of the child whose profile is unlike that of most other children of the same age, feels less anxious because they now have a term which helps them predict what happens next. After all, if a teacher knows that 'David' has a Moderate General Learning Disability, he or she can predict that David will take longer to learn a task or fact and will be less anxious when this occurs.

The problem with labels is that they can become self-fulfilling prophecies. Labels create expectations and expectations are communicated to children through non-verbal behaviours (rolling the eyes when an incorrect answer is given, or not asking a child a question in class). Children quickly learn about the expectations adults have about them. Any good observer of children knows that even the so-called 'thickest' child in the classroom is 'gifted' in knowing the teacher doesn't expect much from him!

This negative side of labelling has been confirmed in a number of interesting studies in which classroom teachers viewed videotapes of children who demonstrated normal behaviour and then watched the same children after being told they are emotionally disturbed. The teachers actually identified behaviours they *thought* were those of disturbed behaviour, once they were *told* the children were disturbed. Teachers seem to consistently rate children lower in terms of social and academic potential when those children have been labelled. Yes, labels can be harmful.

There is considerable debate about labelling in special education. Many argue that labels are helpful because they allow children with special needs to receive special help. Others argue the opposite and say that labels limit opportunities and carry with them a life-long stigma. Deciding whether labels are good or bad is not easy. Many factors are in place.

However, it is undeniable that the behaviours of individuals who are different from others, and the reactions to those behaviours on the part of the others they are different from, are often a function of the label.

What can we do about labels? For one thing we aren't going to be able to eliminate them for quite some time. They are part and parcel of the special education system in Ireland, embedded even in the term 'special' itself. After all, when it comes to special education everyone knows just whom we are talking about. There are a few things we can do however to eliminate the negative impact of labels as outlined by Ysseldyke & Algozzine (1995).*

- Avoid phraseology that emphasises the condition over the people, such as "the autistic", "the learning disabled", "the deaf".
- Speak of the person first, not the disability. Say "children with ADHD", not "ADHD children".
- Don't use terminology such as "handicapped"; remember, a handicap is a barrier imposed by society or the environment.
- Don't sensationalise the conditions by saying things such as "afflicted with", "suffering from" and "victim of" etc.
- Avoid medical terminology such as "patient", "cases", "symptoms".
- Focus on ability, what the child can do; talk about what is possible rather than what is not possible.
- Children with special education conditions do not need to be patronised. They are not "God's little angels" or, as one teacher said to me recently, "Our little special ed girl".

Diversity in Irish Education

Zong, Mbogo, and Dzredic are in third class in school in Dublin's inner city. One is from China, one from Kenya, one from Bulgaria. Each of these children have been in Ireland for less than six months. In the same classroom are Daniel, Patricia, and Robert. One has no living parents and is cared for by his grandmother, one lives in a room with her single

* Ysseldyke, J. E. & Algozzine, B. (1995). Special Education, A practical approach for teachers. Boston: Houghton Mifflin Company.

Part Four

teenage mother, and one visits his father in prison every Sunday. In the same classroom are John, Breda, and Maeve. One comes from a family living in an apartment that cost nearly €500,000, one lives with his drug addicted, violent father and mother, one lives in a council house his parents purchased four years ago. The class teacher was raised on a farm in Kerry, had nine brothers and one sister, mother worked in the house cooking and cleaning, father raised a herd of 55 cattle and 150 sheep, mother and grandmother were teachers, two brothers are teachers also. This is the face of diversity in contemporary Ireland.

Each of these children has been acculturated (the process of acquiring the culture one is raised in) differently. Each of these children brings different cultures to the classroom. Each of these children's culture is vastly different from that of the class teacher. All over the country there are increasing levels of diversity in our schools. Difference in race, ethnicity, language and culture can affect school performance. More and more, as a result of diversity, we must create classrooms that honour and celebrate it, rather than ignore or resist it.

Let's take a look at the classroom in which these children are learning. It is a lovely classroom in a lovely school. All the displays are neatly mounted. In the spring there are daffodils painted by the children; in March the St. Patrick's badges have been made, on Mother's Day, cards have been made and brought home. All the displays are written in English, the teacher says good morning to the children in English, words are in the English script. When it is time for a field trip, the children go to a farm and feed the animals, Irish animals.

But wouldn't it be great if the class were shown how to benefit from their own cultural diversity. Why would the teacher not get the class to learn how to say some phrases in Cantonese, using Zong, in Kiswahili using Mbogo, in Bulgarian using Dzredi? This way, the 'different' children can help to add fun, interest and new learning to the classroom and to their peers; they will feel less 'outside' the cultural majority. The modern classroom/school must be transformed into a place where every child and every teacher's cultural background, their acculturation, is recognised, shared, and taught for the benefit of all. Banks* (1989) has identified four goals of what he refers to as multicultural education:

1. to increase the academic achievement of all students
2. to help all students develop more positive attitudes toward different cultural, racial, ethnic and religious groups
3. to help students from historically victimised groups develop confidence in their ability to succeed academically and to influence societal institutions
4. to help all students learn to consider the perspective of other groups

This sort of education involves attitude changes for all; acceptance of difference, tolerance of difference and crucially, respect of difference. What am I saying? The more different we are made to feel, the more apart from everyone else, the more likely we are to have learning problems. The same logic applies to children who are 'different' because they have a special education need. By embracing them, catering for them and celebrating their achievements (each little one), the whole class will derive great benefit and it should not in any way delay the progression of the whole class, through the curriculum.

Part Four

Using the Curriculum

Although what follows is of particular relevance for teachers, I believe it is extremely important for parents of children with special education needs to become aware of what represents 'best international practice' in inclusive education. In Ireland, more and more children with special needs are being taught in mainstream classrooms. We are fortunate to have a child-friendly curriculum that incorporates many of the teaching strategies recommended for children with special needs as part of its philosophical base. Parents need to be aware of how this curriculum can be used to assist their children in learning more effectively in the mainstream classroom.

---

* Banks, J. A. (1989). Multicultural education: Characteristics and goals. In J. A. Banks & D. A. M. Banks (Eds), *Multicultural education: Issues and perspectives* (2nd ed.). Boston: Allyn & Bacon.

For this reason I include this information here, rather than setting it aside in a special section for teachers. The term 'differentiation' (or "curriculum differentiation") is one that is now central to special education. In the most simplistic terms it means that every child, regardless of ability or 'special need' has an entitlement to access the entire school curriculum and that this curriculum must be adapted by mainstream and specialist teachers to enable the child to effectively access it.

The Revised Primary School Curriculum, 1999, is a child-centred document, which is based on activity-based learning, mixed ability grouping, creativity, exploration, and inclusive education of children with special needs. The curriculum contains some new content such as Drama and SPHE (Social, Personal, and Health Education) that have potential to improve the educational opportunities of children in the mainstream classroom. In Ireland, we have one primary school curriculum which every child in the State has a constitutional right to access. Taking a look at how we can use this curriculum to the advantage of *all* children is central to reaching the goal of inclusive special education. I might refer you at this point to a useful reference for all parents of primary school children, "The Essential Parents' Guide to the Primary School Years", by Brian Gilsenan, also published by PrimaryABC (www.primaryABC.ie).

Curriculum Differentiation – Important for all Children

The concept of curriculum differentiation is central to the special education process. At its most simple and uncomplicated, differentiation is the process of changing things a little within the classroom, to ensure that all children have equal opportunities. Since the mainstream teacher has, according to the Department of Education and Science, the primary responsibility for the education of all children in his or her class, knowing how to go about differentiating their teaching is imperative. **Differentiation is more than just doing things differently; it is a philosophy of teaching that must filter down to the heart and soul of every classroom teacher. It is what good teaching, appropriate teaching, is all about.**

Tomlinson (2000)* states that differentiation is based on this set of beliefs:
- Students who are the same age, differ in their readiness to learn, their interests, their styles of learning, their experiences and their life circumstances
- The differences in students are significant enough to make a major impact on what students need to learn, the pace at which they need to learn it, and the support they need from teachers and others to learn it well
- Students will learn best when supportive adults push them slightly beyond where they can work without assistance
- Students will learn best when they can make a connection between the curriculum and their interests and life experiences
- Students will learn best when learning opportunities are natural

A teacher who shares these beliefs refuses to offer a one-size-fits-all method of teaching. Instead, he or she recognises that children must have choices about what to learn and how to learn it, that children have a right to take part in the setting of their own learning goals, and that the classroom must connect with the experiences and interests of each child.

Working from this child-centred philosophical base, and recognising that there are children with special education needs or learning difficulties in every classroom, the skilled practitioner will differentiate the curriculum in a number of ways, not willy-nilly, but with a clear focus on what is best for each child. Among the ways differentiation can proceed are:

Differentiating the Learning Environment
- Make it student-centred
- Encourage independence
- Be open to new people, materials, and new connections
- Be accepting of others' ideas and opinions before evaluating them

* Tomlinson, C. A. Reconcilable Differences? Standards-based teaching and differentiation. How to differentiate instruction. Vol. 58, No. 1. pp6-11.

Part Four

- Make it complex with a variety of resources, media, ideas, methods and tasks
- Make it highly mobile by encouraging learning outside the walls of the classroom

Differentiating the Content

- Shift content from facts, definitions and descriptions to concepts, relationships and generalisations
- Keep the complexity varied by shifting to inter-relationships rather than considering facts separately
- Provide a variety of material and expand it from the curriculum to the outside world of work, play, and knowledge
- Include the study of people and how they reach their goals and react to obstacles
- Give children an opportunity to learn by the same methods experts learn

Differentiating the Process

- Involve cognitive challenges when appropriate; provide logical problems, critical thinking, creative thinking
- Involve the imagination and intuitive approaches and brainstorming techniques
- Encourage risk-taking and response by stressing that sometimes there is no one right answer
- Provide for group interaction, sometimes grouping children by ability, sometimes by mixed ability
- Encourage collaborative and cooperative group work
- Vary the pace of learning. Some children will move from lower-order thinking to higher-order thinking slower than others
- Accommodate different learning styles, e.g. the visual learner, the auditory learner, the hands-on learner
- Encourage children to question one another, to articulate their reasoning and conclusions
- Provide freedom of choice on topics, methods, products and the learning environment

Differentiate the Product
- Provide real problems that are relevant to the lives of the children
- Use real audiences appropriate to the product; other children, groups of children, other teachers, the principal, parents, or members of special interest groups
- Give real deadlines and encourage and teach time-management skills
- Provide for original manipulation of information rather than regurgitation
- Use appropriate evaluative methods; some children need to talk about what they have learned, some need to write about it, some need to show it by demonstration

Management Strategies for Curriculum Differentiation
- Let children negotiate what they will learn when it's possible to do so
- Provide regular conferences between teacher and students, or groups of students
- Use appropriate grouping strategies and appropriate group work challenges

Curriculum Differentiation for Children with Special Education Needs

Differentiation for children with SEN requires a more focused level of thinking on the part of the teacher. Specific strategies must be considered, taking into account for whom they are most appropriate. Methods of doing this are as varied as the teacher's imagination but some of Tomlinson's examples may prove helpful.

Learning Environment
- Provide places in the room for children to work quietly
- Provide material reflecting a variety of cultures and home environments
- Set clear guidelines for independent work that matches individual needs
- Develop routines that permit a child to get help when teacher is busy with others
- Help children learn that some must move around to learn while others do best sitting quietly

Part Four

Content

- Use reading materials at varying levels of readability
- Put some text materials on tape for those who are the weakest readers
- Provide spelling and vocabulary lists at the reading level of all children, even the weakest readers and spellers
- Present material through auditory and visual means
- Use reading 'buddies' or peered reading every week
- Meet with small groups of children to re-teach an idea or skill for the struggling learner
- Provide opportunities for older children in the school to tutor younger children or weak learners
- Meet with groups of advanced learners to push them higher and higher

Process

- Use tiered activities that permit all children to work with the important skills but proceed with different levels of support, challenge, or complexity
- Provide interesting learning centres that encourage children to explore subsets of a topic of particular interest to them
- Provide personal agenda tasks written by the teacher which contain work for all children that addresses the needs of the weakest and the brightest
- Provide manipulatives and hands-on learning experiences for those who need them
- Vary the length of time a child needs to complete a task to support the struggling learner or extend the advanced learner

Products

- Provide options about how to demonstrate learning, e.g. puppet shows, letter writing, creation of mural, plastic art displays or other visuals
- Allow children to work alone or in small groups on their projects
- Encourage children to create their own product assignments as long as they contain the required elements

With the child-centred curriculum in place in our primary schools, there is ample opportunity for every teacher to differentiate. When there is consistent differentiation in a classroom, the number of children requiring special education services from teachers apart from the classroom teacher, will be reduced.

Thinking creatively in the mainstream classroom

Besides differentiating the curriculum in the ways highlighted above, there are ample opportunities for teachers to support the learning needs of the weakest and brightest children in the classroom by utilising some of the more innovative content areas.

The Creative Arts

The curriculum requires the teaching of the creative arts to all children. The creative arts include the visual and plastic arts, drama, dance, music and storytelling. It is well known that the creative arts open doors to learning for those children who are frequently poorly motivated because of learning difficulties, special education needs, social-economic disadvantage or who live in distressing personal circumstances. Nothing so touches the heart of a child like an opportunity to learn and consolidate what is being learned through the creative arts.

Drama

The drama curriculum is one of the most innovative content areas in the primary curriculum. Drama in the curriculum is not about creating an exciting Christmas pageant or end-of-year show. It is about using drama, storytelling, mime and activity to solidify learning, demonstrate in non-traditional ways what has been learned, and extend learning through creative, dramatic experiences. In the primary and secondary classroom, even those children who are having the most difficulty reading can participate in a dramatic activity; they can question a 'historical character' who is in role, discuss with a contemporary 'scientist', mime the solution to a social problem and so on. They can explore alternative solutions to problems. Drama is fun, it allows movement, everyone can participate whether in wheelchair or unable to see or speak, and it is creative. Drama can be quiet or noisy, passive and watching or active and participating; drama is learning through doing and watching.

The Visual Arts

All children love to draw but the visual arts in the curriculum is not about drawing only. It is about exploring fabric, texture, different media, drawing from life, drawing from the great masters, appreciating the visual and plastic arts, and manipulating concrete materials. The visual arts permit children to demonstrate what they have learned by representing it visually. The visual arts in the curriculum stimulate the creative imagination of all children and there is not one child in Ireland who cannot participate in the visual arts curriculum one way or another. Children can explore and master historical content by re-creating their own unique historical art forms, learning about the materials used to create them, re-creating those materials themselves and using them. Children can explore the world of the visual arts through touch if they do not have sight, through smell if they can not touch or see, through movement around objects of sculpture, and through sensing with all the available senses the smell and feel of fabrics and textures.

Dance

Dance is movement, dance is learning in motion. Dance can be slow or fast; it can be a re-creation of particular steps, or the movement of the body in response to music or sound. The body learns through dance. The brain is stimulated by dance. Dance is rhythmic and patterned and many children with special education needs require rhythmic and patterned experiences. Dance improves memory as the body learns the steps in a dance routine. Dance is about cooperation in movement and collaboration in choreography. Dance is about multicultural exploration when we dance to the music and movement of other countries. Dance permits children to show with pride their unique cultural and social background.

Music

Music and movement, or music alone, provides rhythmic, patterned sets of experiences that permit children to develop motor skills, listening and concentration skills, social skills and language skills. Musical notation is a form of written language. The creation of simple musical instruments permits children who do not want to sing to participate in the musical experience. Listening to music fosters analytical skills as children are required to comment on the musical form, structure, tonality, phrasing, and pulse of the piece. Children with developmental coordination problems can benefit from movement to music as they express themselves with scarves, balloons or other small objects that they must manipulate.

Physical Education

I argue that PE is as much an art form as it is a sports form. I am also of the opinion that far too much of what passes for PE in education is actually sports. Indeed, most of it is highly competitive. 'Real' PE provides opportunities for children to play cooperative games where all are winners, because the object of the game isn't about winning at all but about cooperation to achieve a group objective. PE permits children to move slowly (for those children who are overly active). PE permits children to move fast (for those who move too slowly). PE can be engaged in outdoors or indoors. It can involve the movement of the large or fine muscles, small objects or large objects. There is a place for sport in the life of the school but when PE is confined to competitive sport, nearly everyone ends up losing.

Another important use for physical education is sometimes referred to as 'Adaptive PE'. Physical education activities can, and indeed must be adapted for some children who have motor control, coordination or movement difficulties. Children who have difficulty balancing or moving fluidly can benefit greatly from obstacle courses that can be set up in the classroom, hall or yard. Having an opportunity to crawl over, under, through and around objects; to balance on a beam, to walk like a crab, to jump on one foot, to pull oneself up by the arms all contribute to motor coordination and skill. Many of the activities undertaken by occupational therapists or physiotherapists can be incorporated into PE time for the enjoyment and benefit of all the children, not just those with a special education condition involving motor movement.

The Arts are Healing

For those children with emotional and behavioural disturbances, the creative arts offer a healing opportunity. Many of the children we have identified as emotional or behaviourally 'disturbed' have had a childhood filled with chaotic, disruptive, neglectful or terrorising experiences. The absence of repetitive, patterned, enriching experience has a negative impact on the child's brain. Being patterned in different ways from the brains of those children who have had more normal life experiences, these children tend to be overly active or withdrawn, overly sensitive to criticism, and to sometimes act out aggressively.

Brain growth is sequential and starts at the bottom of the brain; there are specific activities that address each part of the brain, as follows:

- The first part of the brain to grow and develop is the brainstem, responsible for arousal, sleep and fear 'states'. Rhythm, touch and massage can re-pattern this brain region
- The next part of the brain to grow is the mid-brain, responsible for motor regulation and multiple sensory input functions. Its primary developmental goal is motor control and affiliation (our ability to get along with others). Music, movement, and simple narrative experiences facilitate the healing of this brain region
- The limbic system of the brain grows and develops next. It is the part of the brain with the end goal of emotional regulation, empathy, and attachment. Dance, play therapy, art therapy, and nature discovery, including the care of animals in the classroom heals this part of the brain
- The last part of the brain to mature is the cortex, that uniquely human part of the brain that controls abstract cognitive functions and social emotional regulation. Drama and exposure to the performing arts help heal this part of the brain

A Classroom Without Walls

There is no subject in the primary curriculum that cannot be taught outdoors as well as indoors. The poverty of our imagination is all that prevents us from utilising the external environment to its maximum power and effect. Indeed, for children with special education needs, sitting in a classroom all day long can be a form of torture. Some need to get up and move, some find it less than stimulating to sit and listen or work quietly at their seat; others flourish but could be extended more if there were opportunities to bring the content and the methods of learning outside.

A classroom without walls is one in which the teacher extends learning opportunities to the 'real' world outside the schoolroom. Mathematics is taught by creating math's trails around the school or community in which shapes are found, numbers are added, distances estimated, and things are measured with both standard and non-standard measurement. History is learned by talking to older people in the community. Industry and commerce are studied on site in shops and businesses. Traffic counts are taken to study transport safety. Literature is explored while we walk the streets where the characters in our story lived out their fictional or real lives. Sounds are recorded and studied, smells are investigated, insects are collected and categorised, and orienteering is undertaken with a compass and map. In other words, the external environment provides an opportunity for every child, regardless of ability to extend learning into the world in which they live.

Conclusion

The curriculum, whether primary or secondary, is something to be made real, to be experienced as much as taught, in the world outside the classroom. It needs to be adapted to the learning style of every child. It contains opportunities for creative thinking, problem solving, anger management, social skill development and the mastery of all subjects. Children with special education needs do not have to be removed from their peers and taught in small rooms by specialist teachers only. Once we begin to think beyond the boundaries of the curriculum we open ourselves as educators and parents to the multitude of ways in which all children can learn.

Part Four

PART FIVE

Special Education in Ireland: Legislation, Guidelines and Structures

This chapter will present an overview of the most relevant legislation, Departmental guidelines and structures related to special education. It is current at the time of publication but the reader is advised to frequently visit the relevant websites (see Appendix Page 263) related to special education provision, as policies are changing all the time. Each relevant document is outlined in general, some in more detail than others. It is not possible within the scope of this book to present each piece of legislation in detail; for that the reader is referred to the Acts themselves, available from the Government publications office (Molesworth Street, Dublin 2) and on the Internet.

Special Education in Ireland – A Short History

Special education in Ireland. Suffice it to say that it has come from a point, (as was the case in mainstream education), whereby it was nearly entirely in the hands of the Religious, from the founding of the State until the early 1990's. As a result, the Department of Education and Science (DES) had little need to concern itself with policy development and the Government had little need to concern itself with legislation. With the subsequent decline in the numbers of Religious in education, more and more responsibility was placed in the hands of DES officials and other Governmental Departments sharing responsibility for children and their education.

Alongside this came the rapid changes in technology, bringing with it access to sources of information, specifically the internet, for parents and teachers alike. It was soon recognised that the educational provision for children with special needs in Ireland fell badly behind that of other countries. The result has been a steep increase in the amount of activity, both legislative and structural.

Ireland established its National Education System in 1831 with attendance mandatory for children between the ages of six and fourteen. However, as early as 1816, an inspection of publicly funded schools by professional inspectors had begun. By 1892, one hundred and fifty days school attendance was compulsory for all children. The Department of Education was established in 1924. The education of children with special needs alongside their so-called "non-disabled" peers was not considered appropriate and in 1947, St. Vincent's Home for Mentally Defective Children was established and was recognised by the State as an official school. The belief at that time in Ireland, and elsewhere, was that the needs of "special" children were a medical, not an educational issue. In Ireland it was the County Clinics that were responsible for assessing children with mental handicaps. After assessment, the options were usually institutional care or some type of basic training.

Parental pressure and the involvement of the Religious communities resulted in a societal push to provide better services and in 1959, the first Inspector for special education was appointed by the Department of Education.

By 1960 more provision for "mentally handicapped children" was announced by the Minister for Education and the training of teachers and psychologists to work with hearing impaired children was undertaken in the University sector. These initiatives had the support of the Irish National Teachers' Organisation (INTO). Throughout the 1960's, more special schools for children with mental handicaps and also physical/sensory handicaps became recognised by the State. In 1965 the development of specialist schools was supported by the Report of the Commission of Inquiry on Mental Handicap. This report also suggested that in some cases, mainstream schools should include special classes for children who were "slow learners".

In 1965 the 'Report of the Commission of Inquiry on Mental Handicap' was published. Then in 1972 there was a report entitled 'The Education of Children who are Handicapped by Impaired Hearing'.

Throughout the 70's and 80's, the policy relating to the education of children with special needs was the establishment of special schools. Despite the fact that many of these schools were not purpose built and occupied buildings that formally housed institutions, the staff were dedicated, hard working and often highly skilled. The leadership of these schools was especially visionary and included many who today are considered to be in the forefront of the development of Ireland's special education system. Along with this development of special schools came a gradual recognition of the differences between children who were slow to learn and those who had physical or mental deficiencies. At the Colleges of Education, courses for teachers who wanted to specialise in remedial education started developing.

In 1982 a report entitled 'The Education of Physically Handicapped Children' was released, followed by a report called 'The Education and Training of Severely and Profoundly Mentally Handicapped Children in Ireland' in 1983.

In the mid-80's, the worldwide push for integration of children with special needs in mainstream schools began to have an influence on Irish educational policy. Classes for children with special needs began to be established in mainstream schools and more teachers were being educated to cater to their needs. By 1993 over two thousand children were being educated in such classes. It was in 1993 that a major publication exerted influence on special education. The Report of the Special Education Review Committee (SERC) dealt comprehensively with the educational implications of special needs.

It was initiated in 1991, when the Minister for Education appointed a group of recognised experts in Irish special education to review services and make recommendations. It provided a report that reviewed services from pre-school to secondary education, from referral to assessment and intervention. It defined students with special education needs as: "...those whose disabilities and/or circumstances prevent or hinder them from benefiting adequately from the education which is normally provided for students of the same age, or for whom the education which can generally be provided in the mainstream classroom is not sufficiently challenging". The committee favoured as much integration as possible that "...is appropriate and feasible with as little segregation as necessary".

Importantly, the report recommended the establishment of a continuum of educational provision to meet a continuum of special education needs that would allow for:

a) Full-time placement in a mainstream class with additional support
b) Part-time or full-time placement in a special class or school
c) Full-time placement in a residential special school
d) Part-time placement in a Child Education and Development Centre (CEDC) or special school

The report identified key gaps in the provision of special education including gaps in curriculum development, constraints at primary level, drop-out at post-primary level, insufficient specialist training for teachers, and the lack of contact between mainstream and special education systems. Among its final findings, the SERC report noted that the system of special education in place at the time (1993) "...inhibits the realisation of one of the main goals of education for such students (students with special education needs), namely that they should be capable of living, socialising and working in their communities".

In the context of Irish special education, the SERC report was a groundbreaking document. Although it did not meet an appropriate international best practice standard, it certainly provided a comprehensive re-visioning of special education in Ireland and continues to have an impact.

Its influence may first be perceived in the 1995 Government White Paper on Education 'Charting our Educational Future' where it is stated "...all students, regardless of their personal circumstances, have a right of access to and participation in the education system, according to their potential and ability". This White Paper also stated that the objective would be:

"...to ensure a continuum of provision for special educational needs, ranging from occasional help within the ordinary school to full-time education in a special school or unit, with students being enabled to move as necessary and practicable from one type of provision to another. Educational provision will be flexible to allow for students with different needs, at various stages in their progress through the education system."

A key aim stated in the White Paper was "...to promote quality and equality for all, including those who are disadvantaged through economic, social, physical and mental factors, in the development of their full educational potential". In 1996 a report entitled 'A Strategy for Equality; Report of the Commission on the Status of People with Disabilities' was published. It noted the lack of cooperation between special and mainstream schools and the lack of support services for children with special needs. It also cited the lack of a flexible curriculum, lack of transport, and lack of necessary resources and equipment in special schools and classrooms.

In 1998 the first legislation dealing specifically with education in Ireland was passed, 'The Education Act, 1998'. This was followed by a series of legislative initiatives
- The National Disabilities Authority Act, 1999
- The Education Welfare Act, 2000
- The Equal Status Act, 2000

These are described in detail in the next section, see page 136.

Another White Paper, one concerning adult education, which was written in 2000, set out goals for adult education that included access for all and facilitating the needs of diverse groups, including adults with disabilities. It noted that in order to gain access to education, adult learners with disabilities "...may need additional supports to overcome difficulties that relate specifically to the nature of their disability".

Yet another White Paper, this one concerning Early Childhood Education, was released in 2000. It speaks of the importance of early childhood

Part Five

education (the Cinderella of Irish education) as the foundation of all future learning for all children, including those with special needs. The Paper notes the need for a wide range of improvements in services for children with special education needs in this sector including:
- Parental access to an early education expert
- Resource and visiting teacher supports
- Improved training and skill development in special needs for early years educators
- Appropriate curricular guidelines
- Increased supports for pre-school programmes that enrol children with special needs
- More classes for children with special education needs between the ages of three and four
- The extension of the National Educational Psychological Service (NEPS) to all early childhood schools and all children in these schools who have special education needs

On 26 January 2000 the Minister for Education published a list of seven guiding principles concerning rights and responsibilities within special education. The intent was to make these guiding principles the basis for a national support service. The principles are:
1. **Entitlement:** All children with identified Special Educational Needs (SEN) should have entitlement to quality educational services appropriate to their needs and abilities.
2. **Early Identification of Needs:** Assessment leading to the identification of SEN should be comprehensive and should take place at as early a stage as possible in the child's life.
3. **Promoting Inclusion:** Special education services should promote the inclusion of all children with SEN, regardless of disability. The aim of special education provision should be for children/young people with disabilities to share, with their peers, as complete an educational experience as possible.
4. **Review Progress:** The progress of those with identified SEN should be tracked and reviewed at regular specified intervals and at key junctures in the educational process.

5. **Continually Update Policy:** Policy and practice in the area of special education should be based on consideration of the most up-to-date relevant research and on evidence of best practice both at home and abroad.
6. **Integrated Services:** As most disabilities encompass a continuum of needs, there should be a continuum of special educational provision in relation to each type of disability.
7. **Right of Appeal:** An appeals system should be established to deal with situations where differences of opinion arise, in matters of identification and provision, between professionals, the children/young people and their parents/guardians.

Other reports will inevitably follow; each will speak of the rights and entitlements of children with special needs in Ireland. Despite the issuing of these reports, the provision of special education in Ireland remains problematic and stressful for all concerned. Lack of human resource, especially the availability of speech and language therapists, physiotherapists, educational psychologists and occupational therapists complicates the delivery of services to those who need them. The numbers speak for themselves.

Ireland continues to struggle to implement the aims, goals, and objectives of the SERC report and the White Papers. However, these documents, (especially the SERC report), make it clear there is recognition on the part of officialdom that special education in Ireland requires considerable development. Over the past seven to eight years, legislative developments have made it clear that Irish special education is moving forward, haltingly perhaps, but clearly forward towards appropriate provision.

Statistics available in 2001 report that there were 3,162 national (primary) schools in Ireland. Of these, 476 had classes catering for 9,092 students with special needs and 125 were special schools educating 7,124 students. The rapidly increasing numbers of children with special education needs enrolled in mainstream schools, and the increase in educational provision for children with autism in special units and classes in mainstream schools has placed stress on every aspect of educational provision, from teacher training to classroom practice. The situation at secondary level is troubling at present with little resource having been devoted to that sector. This fact, combined with its exam driven curriculum make it likely that the large numbers of primary students with special needs who will enrol in secondary schools over the coming years will find it enormously difficult to access the level of supports they require and are accustomed to. The greatest challenge facing the special education system will be at secondary level for sure.

Legislation

The Education Act, 1998

In 1998 the Education Act was passed. This was the first piece of legislation that outlined the legal rights and responsibilities of the Government relating to education. This Act is broad and encompasses all aspects of education but certain sections of it have relevance to special education. The most essential of these are outlined below.

The Act provides the fist **legal definition of disability** and states:
" 'Disability' means-

(a) the total or partial loss of a person's bodily or mental functions, including the loss of a part of the person's body,

or

(b) the presence in the body of organisms causing, or likely to cause, chronic disease or illness,

or

(c) the malfunction, malformation or disfigurement of a part of a person's body,

or

(d) a condition or malfunction which results in a person learning differently from a person without the condition or malfunction,

or

(e) a condition, illness or disease which affects a person's thought processes, perception of reality, emotions or judgment, or which results in disturbed behaviour;" (Part 1, Section 2).

The Act **defines a parent** as a birth parent, a foster parent, a guardian appointed under the Guardianship of Children Act 1964 to 1997, any other person acting *in loco parentis* by an order of a court, or a person who has adopted a child or children. (Part 1, Section 2).

In this Act a **school is defined** as "...an establishment which provides primary education to its students and which may also provide early childhood education or provides post-primary education to its students and which may also provide courses in adult, continuing or vocational education or vocational training." (Part 1, Section 2).

The Act creates the first **legal definition of special educational needs**, which it describes as "the educational needs of students who have a disability and the education needs of exceptionally able students." (Part 1, Section 2).

Support services are defined in the Act as "...assessment of students, psychological services, guidance and counselling services, technical aid and equipment, including means of access to schools, adaptations to buildings to facilitate access and transport for students with special needs and their families, provision for students learning through Irish sign language or other sign language, speech therapy services, and provision for early childhood, primary, post-primary, adult or continuing education to students with special needs otherwise than in schools or centres of education." (Part 1 Section 2).

The Act also ensures that the **constitutional rights of children** with special education needs, or who have a disability, will be protected (Part 1, Section 6, a). The reference to constitutional rights refers to the state's obligation to provide for the education of every child. However this citation to a constitutional right is compromised in the following section of the Act, which states every person covered under the Act shall have regard to the following: "...to provide that, as far as is practicable and having regard to the resources available, there is made available to people resident in the State a level and quality of education appropriate to meeting the needs and abilities of those people." (Part 2, Section 6, b).

Part Five

The Act requires the Minister to provide funding for each recognised school and centre for education and to provide support services for students, including those who have a disability or special education need. (Part 1, Section 7, 2, a).

In defining **the functions of a school**, the Act states that a recognised school will ensure that the educational needs of all children will be provided for in a manner that is appropriate to their needs and abilities. It states that the school will "ensure that the educational needs of all students, including those with a disability or other special educational needs, are identified and provided for." (Part 9, a).

Part Three of the Act relates to **the Inspectorate**, which is the division of the Department of Education and Science that has responsibilities for the quality assurance outlined in the Act. There are presently one hundred and fifty inspectors within the primary and post-primary sector. The work of the Inspectorate is managed by the Chief Inspector and organised into two divisions; Regional Services and Policy Support. A Deputy Chief Inspector heads each of these divisions. The Inspectorate evaluates schools and teachers; the Chief Inspector issues reports that can be accessed on the Department's website.

At present a comprehensive school report ("Tuairisc Scoile" in Irish) is completed on each primary and special school about every six to seven years following a comprehensive school inspection. This inspection investigates every aspect of the school's functioning, from teaching to assessment. (It is not yet available to the public under the Freedom of Information Act, but changes in this policy are being considered at time of publication).

The Act makes it clear that the Minister must include persons in the Inspectorate "who hold qualifications as psychologists or who have other expertise, including expertise in the education of students with special educational needs" (Part 3, Section 13, 2). The exact functions of the Inspectorate related to special education are outlined in the Act and include:

- "Consultation with parents to assess the psychological needs of students and to advise the students and their parents and the schools in relation to the educational and psychological development of the children
- To advise schools about policies and strategies for the education of children with special needs
- To advise the Minister about any matter related to the psychological needs of children
- To collaborate with parents and principals and teachers in the creation of "…a school environment which prevents or limits obstacles to learning which students may experience"
- To advise the Minister in relation to the linguistic needs of deaf children in recognised schools" (Part 3, Section 13, a-e).

Part Four of the Act relates to, and requires the establishment of **Boards of Management** and specifies as one of the duties of a Board the necessity of publishing the policy of the school which must include matters relating to "…participation by students with disabilities or who have other special educational needs…" (Part 4, Section 15, d). This section also makes specific the rights of parents to send their children to a school of their choosing.

An important requirement of the Act under this section is the obligation of a Board to publish a **school plan**. The Act specifies that the plan must state the objectives of the school relating to equality of access to, and participation in the school by all students, including those with disabilities or who have other special educational needs, and the measures that the school proposes to take to achieve those objectives. The Act states, "the school plan shall state the objectives of the school relating to equality of access to and participation in the school and the measures which the school proposes to take to achieve those objectives, including equality of access to and participation in the school by students with disabilities or who have other special education needs." (Part 4, Section 21, 2).

Part Five of the Act outlines **the responsibilities of the Principal and teachers** within the school. It states that the principal and teachers shall

Part Five

encourage and foster learning in students, regularly evaluate students and report the results to the parents, and promote cooperation between the school and the community that it serves. (Part 5, Section 22, 2).

Part Six of the Act indicates that a **process of appeal** may be established by the Minister (Part 6, Section 28,1a). This appeal process will be extended to parents or students who have reached the age of eighteen. The appeal process may be applied against the Board of Management of a school, a teacher or any member of staff.

There is also mention that the Minister may make regulations related to "...access to schools and centres for education by students with disabilities or who have other special educational needs, including matters relating to reasonable accommodation and technical aid and equipment for such students;" (Part 6, Section 33, i).

The Act defines one of the **functions of the Minister for Education and Science** as "...to ensure subject to provisions of this Act, that there is made available to each person resident in the State, including a person with a disability or who has other special education needs, support services and a level and quality of education appropriate to meeting the needs and abilities of that person." (Part 7, a). The Minister also has the duty to plan and co-ordinate support services." (Part 7, Section c, 2).

The Minister is also required to "...provide funding...and support services to...students who have a disability of who have other special education needs, and their parents as the Minister considers appropriate..." (Part 7, Section 2, a).

The Act created the framework for the establishment of the **National Council for Curriculum and Assessment (NCCA)**, whose function it is to advise the Minister on all curricular and assessment matters including "...on the requirements as regards curriculum and syllabuses, of students with a disability or other special educational needs;" (Part 7, Section 41, f).

General Impression of the Education Act

The Education Act is a major step forward in Irish educational policy and provision, but a careful reader can easily see its limitations for children with special education needs. The Act includes many vague and ambiguous phrases such as "the Minister *may*", 'the Inspectorate will *advise*', and other such terminology. The single, most critical phraseology releasing the Government from any significant responsibility to children with special needs occurs on page eleven of the published Act and states "In carrying out his or her functions, the Minister "shall have regard to the resources available." (Part 1, Section 7, 4a, (i)).

As all parents and teachers are aware, there are currently insufficient resources available to provide appropriate special education services for many children with special needs. The Education Act is ineffective in its ability to ensure that the educational provision for children with special needs will be the most appropriate provision possible.

Though obviously well intentioned, the Act does not contain imperatives directing the people responsible for special education provision to do what they *must* do. It is largely aspirational in that it speaks of what *should* or *might be done,* were it possible to do so. As a result, it may well be destined to be ineffective legislation, from the perspective of the child with special education needs.

Part Five

The National Disabilities Authority Act, 1999

This Act created the National Disability Authority and defined its functions. The Act defined disability as "...a substantial restriction in the capacity of a person to participate in economic, social, or cultural life on account of an enduring physical, sensory, learning, mental health, or emotional impairment." (Part 1, Section 2).

The functions of the Authority are wide-ranging and relate to all areas of disability including:
- Assist the Minister for Health in coordinating and developing policy for persons with disabilities
- Undertake, commission, and collaborate in research related to disabilities
- Advise the Minister on standards of programmes for people with disabilities
- Monitor the implementation of standards and codes of practice.
- To prepare codes of practice
- To liase with other bodies providing services to people with disabilities (Part 1, Section 8).

Part 2, Section 9 requires the Authority to issue a Strategic Plan every three years. This plan will:
- State the key objectives, outputs and related strategies, including the use of resources, of the Authority
- Be prepared in accordance with the requirements of the Minister for Health
- Have regard to the need to ensure the best use of the Authority's resources

Part 2, Section 14 of the Act gives the power to the Authority to request information from any body providing services to persons with disabilities, with a view to determining the adequacy of the provision. If the Authority determines the provision is not adequate, or not being provided when statutes dictate it should be, the provider will be alerted by the Authority of its failure.

The Authority is composed of a Chairperson and twenty members.

General Impressions of the National Disabilities Authority Act

This Act creates a legal body responsible for overseeing all aspects of service provision to people with disabilities. It also creates the necessary body to see to it that legislation related to people with disabilities is implemented appropriately. The Act is clear and concise and it's a necessary and welcome addition to the entire scope of disability service in Ireland.

The Education Welfare Act, 2000

The main purpose of this Act is to ensure that every child in the State attends a recognised school or otherwise receives an appropriate education. The Act also requires schools to see to it that all children, regardless of disability or special education need, participate in and benefit from the entire life of the school. The Act created the **National Educational Welfare Board**, whose purpose it is to promote and foster, in society and families, recognition of the importance of education in the lives of children.

The Act states the age of **mandatory school attendance** as follows: "…a person resident in the State who has reached the age of six years and who-
(a) has not reached the age of sixteen years, **or**
(b) has not completed three years of post-primary education, whichever occurs later, but shall not include a person who has reached the age of sixteen years" (Part 1, Section 2).

The **functions of this Board**, as outlined in the Act, include the following:
(a) "…to promote and foster in society, and particularly in families, an appreciation of the benefits to be derived from education, in particular as respects the physical, emotional, social, cultural, and moral development of children, and of the social and economic advantages that flow there from
(b) to promote and foster, in recognised schools, an environment that encourages children to attend school and participate fully in the life of the school
(c) to conduct and commission research into the reasons for non-attendance on the part of students and into strategies and programmes designed to prevent it
(f) to advise and assist children and the parents of children who exhibit problems relating to attendance at, and behaviour in, school" (Part 10, Section 1, a-c, f).

Under this Act, when a child is not attending school regularly, the Board may require an assessment of the child to determine what factors relate to non-attendance. Should the parents or guardians of the child refuse this assessment, the Board may appeal to the Circuit Court and seek an order to assess.

This is outlined in the Act accordingly:
- (4) "The Board may, with the consent of the parent of the child concerned, arrange for a child to be assessed as to his or her intellectual, emotional, and physical development ... by such person as determined by the Board with the concurrence of the parent.
- (5) Where a parent refuses to give his or her consent, under subsection (4), the Board may apply to the Circuit Court for an order that an assessment of the child be carried out" (Part 10, section 4-5).

The Board appoints Educational Welfare Officers to investigate cases of non-attendance and the staff and Boards of Management of schools are required to provide whatever information or assistance is necessary in the course of the investigation.

> "The Board of Management, principal, teachers, and other members of staff of a recognised school shall give all such assistance as may be required by an educational welfare offices in the performance by the educational welfare officer concerned of his or her functions." (Part 11, Section 4).

In the case of parents who are sending their child to a school that is not recognised by the DES, the Act outlines a process whereby they can place the child's name on a register of children not attending a recognised school and seek permission to retain the child in the school of their choice. "The parent of a child who ... is being educated in a place other than a recognised school shall, if he or she wishes the child to be so educated, apply, not later than 3 months after such commencement, to the Board to have the child registered in the register." (Part 14, Section 3).

The Act goes on to specify, in further sections of Part 14, that the Board will assess the child and the educational setting in which the child is enrolled, to determine its suitability. Following this assessment, the Board will notify the parents if continuation in that setting is to be allowed and the child's name be placed on the register. From time to time the Board will assess the suitability of this setting for the child and may conduct additional assessments on the child himself or herself. In the case of a child who is received education in non-recognised school and whose name is not on the official register, the Board may require the parents to enrol the child in a recognised school.

The Act states "The board (Board of Management) of a recognised school shall not refuse to admit as a student in such school a child, in respect of whom an application to be so admitted has been made, except where such refusal is in accordance with the policy of the recognised school concerned published under Section 15(2) of the Act of 1998 (The Education Act)." This refers to a Board of Management's responsibility to publish and make available to parents its policy on the admission of children with special education needs. The Act stipulates that the school principal must retain an attendance register and submit the name of any child who is not a regular attendee to the Welfare Board. The principal is also not to remove a child's name from the attendance register of the school except in certain circumstances of which the Board would be well informed previously.

The Act goes on and states that the Educational Welfare Board "...shall make all reasonable efforts to ensure that provision is made for the continued education of the child and his or her full participation in the school." The Board of Management of a school must submit to the Educational Welfare Board a statement outlining the strategies it intends to take to assure good attendance. It must also report on those aspects of the operation and management of a school, and the teaching of the curriculum, that may result in poor attendance and indicate what strategies will be undertaken to remove these obstacles. Boards of Management must also identify, at an early stage, those students who at risk of developing school attendance problems and create strategies to reduce the likelihood that they do so.

General Impression of the Educational Welfare Act

The implications of the Educational Welfare Act are potentially far-reaching. It is not uncommon for children with social-emotional or behavioural difficulties to be poor school attendees. There may also be a temptation for a school to turn a blind eye on the poor attendance of these children. Additionally, schools may sometimes *cause* poor attendance as a result of not making appropriate educational plans for children with special needs. This Act makes it possible for parents to see to it that a school exhausts every possible means in assessing the causes of poor attendance and that school-related problems are addressed with appropriate solutions.

The requirement under the Act that children with poor attendance must be assessed to ascertain the reason for their absences will no doubt result in the discovery that significant numbers of these children have special education needs. Once a special need has been identified, the school will be obliged to provide appropriate special services to increase attendance and assure that educational provision is modified and tailored to the needs of the child.

The Equal Status Act, 2000

The Equal Status Act, 2000 promotes equality of opportunity for all citizens and legal residents of the state. It prohibits discrimination on nine grounds:
1. Gender
2. Marital status
3. Family status
4. Sexual orientation
5. Religion
6. Age (for persons over 18)
7. Disability

8. Race
9. Membership of the Travelling Community

The definition of each ground of discrimination is as follows:
1. Gender – being male or female
2. Marital Status – being single, married, separated, divorced, or widowed
3. Family Status – being pregnant or having responsibility as a parent in relation to a person under 18 years, or as a parent or the resident primary carer in relationship to an adult with a disability who needs care or support on a continuing, regular, or frequent basis
4. Sexual Orientation – being heterosexual, homosexual, or bisexual
5. Religion – having religious beliefs or having none; the term "religious beliefs" includes religious background or outlook
6. Age – applies to everybody over 18
7. Disability – the term "disability" is broadly defined. It covers a wide range of impairments and illnesses. It covers all physical, intellectual, and sensory disabilities
8. Race – includes race, colour, nationality or ethnic or national origins
9. Membership of the Travelling Community – being a Traveller

This Act has particular relevance to special education in a number of ways. Firstly, it prohibits three forms of discrimination: direct, indirect, and discrimination by association:
- Direct discrimination occurs when a person is treated less favourably than another person on the basis of one of the above nine grounds
- Indirect discrimination occurs when a person must comply with a requirement or practice and cannot do so because of one of the defined grounds
- Discrimination by association occurs when a person is treated less favourably because he or she is associated with a person falling under protection based on the nine grounds

Under this Act, no member of a school staff may harass a student, or anyone applying for admission to a school, under any of the nine grounds. The Act requires that schools provide "reasonable accommodation" to students with disabilities or special education conditions. Reasonable accommodation is defined as making adjustments to permit a person to access a service, in this case the educational services of the school.

The Equal Status Act specifies four areas in which a school must not discriminate:
- Admission
- Access
- Any other conditions on participation in the school (e.g. rules restricting permission to take examinations)
- Expulsion

Certain exemptions to these requirements apply. For example, admission to a single-sex school is restricted to members of the same sex. A school can refuse admission on religious grounds, to maintain the religious ethos of the school. However, there is no exemption related to disability or special education condition.

Denial of access to courses, facilities and benefits is prohibited under this Act and means that any student enrolled in a school is entitled to access the full life of the school, including extra-curricular activities such as school trips, assemblies and other functions.

Interestingly, expulsion from school, as a result of any of the nine grounds for discrimination, is prohibited. This could have repercussions for those children who are suspended from school for exhibiting behaviour that is a direct result of their special education condition. Such a claim would have to take into account the need to protect all children and students from harm, because this requirement is stipulated in both the Education Act and in this Act.

The foundation upon which the Equal Status Act rests in relation to special education is 'mainstreaming'. The only exemption to this requirement is if the nature of the student's disability makes it impossible to educate other students, or it would present a seriously detrimental effect on the education of other students.

The Act underscores the importance of the school plan and indicates that it must specify the school's equality objectives and specify the steps that will be taken to achieve them. Additionally, the Act clearly indicates that admission to a school must not be refused based on any of the nine grounds of discrimination. The school's code of behaviour must also address discrimination issues. It should identify actions that will be taken to prevent discrimination on all the nine grounds and what steps will be taken to deal with harassment on any of the grounds.

General Impressions of the Equal Status Act

This Act has significant implications for children with special education needs. It is based on the premise that children with special needs should be educated in the mainstream in so far as is possible. It has a clear implication that a child who is expelled from school may be being discriminated against if the reasons for expulsion arise from a special education condition. It implies that it is discriminatory if a child is excluded from a school activity as a result of circumstances or behaviours arising from a special education condition. The Act may provide protection for children whose rights are only vaguely protected under other educational legislation.

The Children's Act, 2001

This Act relates mostly to the protection of children, the legal rights of children, and to children who have been involved in juvenile offences. The Act defines a child as a person under the age of 18. Part 2, Section 7 of the Act defines that a Family Welfare Conference may be convened upon order of the Children's Court, if the Health Executive comes to the opinion that such a Conference is necessary to ensure the proper care or protection of a child. A Family Welfare Conference may decide if a child is in need of special care or protection and recommend to the Health Executive that an order to provide care and protection be sought from the Courts. The following people are entitled to attend a Family Welfare Conference:

- The child
- The parent(s) or guardian(s)
- Any guardian *ad litem* (a person appointed by the Court to represent the rights and interests (including educational interests) of a child) appointed for the child
- Other relatives appointed by the Conference Coordinator after consultation with the parents/guardians and the child
- An officer or officers of the Health Executive
- Any other person who the Conference Coordinator feels, after consultation with the child and parent(s)/guardian(s), may make a contribution to the conference (Part 2, Section 9).

If the courts are asked to intervene as a result of a Family Welfare Conference, they may make a special care order if the child's behaviour poses a risk to his or her health or development, or if the child requires special care and protection he or she is unlikely to receive unless an order is made. (Part 3, Section 16).

The Act goes on to describe a wide range of child protection rights, including the rights of children who have violated the law.

General Impressions of the Children's Act

This Act provides important protection to children having needs for special care or protection. It might serve as a protective measure to ensure that children who have special education needs receive all the services they require.

The Teaching Council Act, 2001

The purpose of the Teaching Council Act is to regulate the teaching profession at primary level. It oversees the professional conduct of teachers. It will also establish standards in teacher education and all other levels of the professional education of teachers. The Council will establish and maintain a Register of Teachers. All primary school teachers will eventually have to be registered with the Council. This will apply to all primary teachers, whether or not they are currently employed in teaching positions. Registration will have to be renewed every year. The Council will also advise the Minister for Education and Science on teacher supply, entry qualifications for the profession and teacher professional development. There are currently thirty-seven members on the Teaching Council.

The exact functions of the Teaching Council are still in the planning stages. However, one of its primary functions will be to ensure that all teachers maintain membership on the register. Towards this purpose, it may remove a teacher from the Register on certain grounds, such as professional misconduct, fraudulent registration or medical unfitness. The Council will also put into effect a process for induction and mentoring of newly qualified teachers.

The Act creates an investigating committee that will look into all allegations related to fitness to teach. A complainant may be the Council itself or any person who has reason to make a complaint. Under the terms of this Act, failing to comply with, or contravening any aspect of this Act, or the Education Act of 1998, or the Education Welfare Act 2000 are grounds for a complaint (the same may be true in relation to the Education of Persons with Special Education Needs Act). Following investigation, the committee may refer the case, in whole or in part, to the disciplinary committee for final determination of the outcome. Sanctions, if the complaint is upheld, may involve suspension from teaching or removal from the teaching register.

The Teaching Council is in place at the time of writing, but its respective duties and functions are not yet entirely determined. It remains to be seen what impact the Council will have in cases where parents believe a teacher to be unfit as a result of not appropriately catering for the educational needs of children with disabilities.

General Impression of the Teaching Council Act

This Act could have significant impact on the rights and entitlements of children with special education needs. It makes it mandatory for teachers to provide professional levels of skills to all children. Under this requirement, it may be necessary for teachers to keep current with special education strategies in mainstream classrooms. This might require teachers to engage in life-long learning on either a formal or informal basis. It might put an end to the excuse used by some that since they had no training or education in special needs, they don't know how to help these children. Parents may have a right to appeal in such cases, on the grounds of professional malpractice.

The Equality Act, 2004

This Act provides additional protection in the workplace against harassment and unfair treatment. It makes specific mention to the rights of people with disabilities in the workplace. It outlines the duties of an employer of a person with a disability to take appropriate measures as follows:
- To enable a person with disability to have employment
- To participate or advance in employment
- To undergo training (unless the measures would place an undue burden on the employer) (Part 2, Section 9).

"Appropriate measures" are defined in this Act as:
- Effective and practical measures to adapt the workplace to the disability concerned
- Adapting the premises and equipment, patterns of working time, distribution of tasks or the provision or training or integration resources

It does not include any treatment, facility or thing the person himself or herself might reasonably provide themselves (Part 2, Section 9).

General Impression of the Equality Act

Although not directly related to children, the Act further outlines the rights of the disabled in Ireland. The reader should take note of the usual escape from responsibility clause noted in the terminology "undue burden".

The Education for Persons With Special Education Needs Act 2004 (EPSEN)

This is the most significant piece of legislation in the history of the State related to the education of children with special needs. Passed into law in July 2004 and becoming fully implemented no later than January 2009, this Act defines the entire scope of special education provision. *Because this Act is of such critical importance it will be reviewed here without the large number of page citations found in the review of other Acts. Parents, teachers, and all concerned with the education of children with special needs should become familiar with the entire contents of this Act as a matter of urgency.*

The Act creates several new bodies with duties and responsibilities for special education. These are the National Council for Special Education and the Special Education Appeals Board.

The Act focuses on:
- A definition of special education needs
- The nature of assessment
- The content of the educational plan
- The designation of schools for children with special needs
- The duty of all schools related to children with special needs

Generally, the Act provides for the education of all children, including those with special education needs, in an inclusive environment in which they will:
- Avail of, and benefit from, appropriate education
- Acquire skills to participate to the level of their capacity
- Permit greater involvement by parents in the special education process

The Act speaks of the type of **inclusive educational environment**; this means that how and where a child will be educated depends on the severity of their special education condition and its impact on the education of other children. The severity of the condition will have to be determined by assessment but regardless of the degree of severity, the educational interests of other children will have to be taken into account when determining where a child with special needs will be educated.

It might be beneficial to begin at the beginning and quote from the opening of the Act itself...

"An Act to make further provision, having regard to the common good and in a manner that is informed by best international practice, for the education of people with special educational needs to provide that the education of people with such needs shall, wherever possible, take place in an inclusive environment with those who do not have such needs, to provide that people with special educational needs shall have the same right to avail of, and benefit from, appropriate education as do their peers who do not have such needs, to assist children with special educational needs to leave school with the skills necessary to participate, to the level of their capacity, in an inclusive way in the social and economic activities of society and to live independent and fulfilled lives, to provide for the greater involvement of parents of children with special educational needs in the education of their children..." (EPSEN Act, preamble).

Lofty aspirations indeed, and noble too! What follows is a closer look at the provisions of this important Act, section by section, followed by reflection about its potential impact.

Section Two of the Act states that a child with a special education need will be educated in an inclusive environment with children who do not have such a need, unless doing so is inconsistent with:

- The best interests of the child as determined by an assessment
  Or
- The effective provision of education to the other children with whom the child is to be educated

Section Three outlines the responsibilities of the **school principal** to take "...such measures as are practicable to meet the educational needs of the student concerned." (Section 3, 2). If these measure are deemed unsuccessful and the student is still, in the opinion of the principal, not benefiting from the educational programme (and the principal believes this is due to a special education condition), the principal, after consultation with the parents, shall arrange for an **assessment** to be undertaken on the child.

The Act specifies that the relevant Health Board (now known as The Health Executive) or the NCSE can request that an assessment is carried out on a student if the Health Executive believes the child has, or may have, a special education need.

The principal may however, with the consent of the SENO (Special Education Needs Organiser, see page 189), decide that any guidelines issued by the NCSE relating to **an educational plan** may not be complied with, provided there are substantial reasons for so doing. If the principal takes the opinion that the plan will not meet the child's special educational needs (if that opinion arises out of a review of an implemented plan), the principal can request the NCSE to prepare another educational plan.

There are timelines for this eventuality; work on the revised plan must be undertaken not later than one month after notification, and must be completed no later than two months from its commencement. Under this contingency, the SENO will convene a **team** to provide advice in relation to the plan. This team will include the parents, the school principal or a teacher nominated by that principal and may include the child (when considered appropriate by the SENO), a psychologist employed by the DES or the National Educational Psychological Service (NEPS), and any other persons the parents or SENO consider appropriate (providing they have relevant qualifications related to the education of children with special needs).

If the NCSE refuses to accept a request to prepare an educational plan, the parents or the principal may **appeal** against this decision to the NCSE Appeals Board (see page 189). The Appeals Board may allow or dismiss the appeal. If allowed, the board may direct the NCSE to arrange for an

Part Five

assessment, or prepare an educational plan. If the board dismisses the appeal, there is no further recourse under this Act.

If the parents of a child who is not currently enrolled in school, because the child is not yet of school age, believe that a child has a special education need, they may request an assessment by the Health Executive or the NCSE. One month from receipt of the request, the Health Executive or NCSE should commence an assessment to be completed without undue delay (but no specific time limit is specified in the Act).

The Health Executive or the NCSE may refuse to perform an assessment if they believe there are insufficient grounds or if an assessment has been completed within the previous twelve months. If either agency refuses a request for assessment, the parents may appeal to the Appeals Board. Within six weeks from the date of an appeal, a decision will be rendered – the Board may require an assessment or dismiss the appeal.

Section Four outlines the **timelines for an assessment** to be completed:
- It shall commence as soon as possible, not later than one month after the Principal has reached the opinion that an assessment is necessary
- It shall be completed as soon as practicable, and not later than three months after the opinion has been reached

Sections Five to Seven define the **guidelines relating to the persons completing the assessment**, describing how they must adhere to the form such assessments are to take, as described by the National Council for Special Education (NCSE), which is described in some detail later in the book (see page 188). These sections cover the Appeals process in relation to assessments and how particular services that are required for a child are to be provided. Since both the Health Executive and the NCSE have shared responsibility for children with special needs, the Act specifies the ways in which they will determine who is responsible for what.

Sections Eight to Twelve of the Act deal with the important area of **the Education Plan.** The Act makes it clear that if a student has a special education need (as identified by assessment), the principal will, within one month of receiving the assessment, set in motion the preparation of an education plan. The preparation of an education plan is under the guidance of the school principal who will ensure that the parents of the child, the Special Education Needs Organiser (see page 189) assigned to the school and any other persons the principal considers appropriate, are consulted. The parents are to have their involvement in the plan's preparation facilitated by the school principal.

An educational plan can be prepared at school level under the direction of the principal and in consultation and cooperation with the SENO and parents **or** it can be prepared under the direction of the NCSE in cooperation and consultation with the SENO. The plan must be prepared no later than one month after the NCSE directs its commencement and must be completed no later than two months after conception.

In the preparation of an educational plan, the SENO convenes a 'team' of people, to advise him or her about the plan's preparation. The team will include the parents of the child and the principal of the relevant school. It may also include a psychologist, another health professional such as a speech and language therapist, a physiotherapist, an occupational therapist, a child psychiatrist, or other representatives of the relevant school (defined as either the school the child attends or the school it is proposed he or she will attend).

An educational plan, according to the Act, must include statements concerning:
- The nature and degree of the child's abilities, skills and talents
- The nature and degree of the child's special educational needs and how those needs affect his or her educational development
- The present level of educational performance of the child
- The special educational needs of the child
- The special education and related support services to be provided to the child to enable the child to benefit from education and to participate in the life of the school

Part Five

- Where appropriate, the special education and related services to be provided to the child to enable the child to effectively make the transition from pre-school education to primary school education
- Where appropriate, the special education and related support services to be provided to the child to enable the child to effectively make the transition from primary school education to post-primary school education, and
- The goals which the child is to achieve over a period not exceeding 12 months

At the time of writing, there is no one format for the writing of educational plans but one is expected shortly from the NCSE. The reader is advised to watch the relevant websites (see page 263) and look out for the format, when it is released.

It is the responsibility of the school principal **to implement the educational plan**. The necessary funds and support services are to be provided by the NCSE in order to do so. If a transfer from one school to another is part of the plan, the principal of the sending school must consult with the principal of the receiving school before a transfer is made. The purpose of this consultation is to ensure continuity of the educational plan and assistance, when required, in any amendment of the plan based on the needs of the child and the structures and operation of the receiving school. The child's parents must be informed by the principal of the receiving school of any such amendments.

The Act speaks directly on the issue of **designating a school for a child with special needs**. If it was the NCSE that directed that assessment and educational planning be undertaken, the NCSE may designate the school the child will attend. Such designation may also be undertaken by the NCSE at the request of the parents. In designating a school, the NCSE must take into account the needs of the child, the wishes of the parents, and the capacity of the school to meet the child's needs. The Board of Management of the designated school may appeal to the Appeals Board within four weeks of being notified of the designation, or if they are concerned about the necessary resources needed to support the designation. The board may allow the appeal and cancel or change the recommendations concerning the designation, or may dismiss the appeal. The burden of proof related to lack of resources is upon the Board of Management of the designated school.

Parents may appeal to the **Appeals Board** if the NCSE refuses to honour their request for a particular school placement. The board may allow or dismiss the appeal. These appeals must be heard and decided within two months of their filing.

Educational plans must be reviewed at least once a year. These reviews will determine if the child has received the services recommended and is achieving the goals specified in the plan. The school principal must make a report to the parents and the relevant SENO concerning the outcome of the review.

If the SENO, upon receiving the principal's report, believes that the goals of the plan have not been achieved, he or she may reconvene the team, or part of the team, and amend the plan.

If parents believe that the educational goals in the plan are not being achieved and a review has not occurred in the past six months, they may request that the principal arrange for a review. If the principal agrees that a review is necessary it will be carried out – if not he will notify the parents in writing within two weeks of having received their request. Within one month of receiving notification from the principal that a review will not be granted, the parents may appeal to the Appeals Board, which must hear the appeal and make a decision within one month of receiving the appeal and may either dismiss or grant a review.

The appeals process may be implemented by parents against actions of the NCSE or a school principal, in relation to any statement or description of their child's special needs that is included in an education plan. Parents may also institute an appeal if they believe any aspect of the plan has not

Part Five

been implemented either by the school or Health Executive. Grounds for an appeal are outlined in the Act and are described in the National Disability Authorities information book about IEP's.* They include the following:

- "Principal and parents may appeal against the Council's refusal to prepare a plan following a request from the principal or health board (Section 3.13)
- Parents may appeal against the Council or health board's refusal to undertake an assessment (Section 4.7)
- Parents may appeal against an assessment on the grounds that it wasn't carried out in accordance with the relevant standards (Section 6.1)
- The board of management of a school may appeal a decision by the Council to designate its school as the recipient of and provider for a specific child with SEN (Section 10.3)
- Parents may appeal against the Council's refusal or failure to designate a school for their child (Section 10.6)
- Parents may appeal against a principal's refusal to arrange a review of their child's education plan (Section 11.6)
- Parents may appeal against the discharge of duties in relation to their child's education plan, by the Council, principal, school or health board (Section 12)"

*National Disability Authority (2005). International Experience in the Provision of Individual Education Plans for Children with Disabilities. Dublin: National Disability Authority. (www.nda.ie)

The Appeals Board must hear and make a determination on these appeals within two months of receiving them. Appeals may be either granted or dismissed.

Section Thirteen to Fifteen outlines the duty of the Minister for Education and Science, and the Minister for Health and Children to provide the resources to facilitate the preparation and implementation of Education Plans. They also deal with the duty of the school and the Board of Management of the school.

Summary of other sections/provisions of EPSEN

Both the Minister for Education and Science and the Minister for Health have duties under this Act. Primarily they must provide whatever funds as may be required to implement an educational plan. Such funding will be under the review and consent of the Minister for Finance as well. The guiding principle of this duty is the constitutional right of all children to receive and benefit from appropriate education.

An interesting aspect of the Act is the power it gives to the Minister for Education and Science to create a 'mediation' process. Where a person brings a complaint directly to the Minister, or intends to take proceedings in the courts alleging that appropriate special education services have not been provided, the Minister may require that the matter be brought before either a person, or persons, to be known as a 'mediator', who will inquire into all aspects of the case. This mediator will receive information from all parties concerned, and will prepare and give a report to all those parties. Mediation cannot be taken if an issue is under appeal to the NCSE Appeals Board. If court proceedings are undertaken, the process of mediation can be entered into and the proceedings can have an impact on the issue of costs being awarded. For example, if it was found that a parent refused mediation, or did not enter it in good faith, the court will have been made aware of this in its decision about awarding costs.

The Act outlines the duty of schools as well as the duty of Ministers. Schools must:

- Ensure that parents are fully informed of their child's needs and how these needs are being met
- Ensure that parents are consulted and invited to participate in the making of all decisions of a significant nature concerning their child's education
- Cooperate to the greatest extent practicable with the NCSE
- Ensure that all relevant teachers and other relevant employees of the school are aware of the special educational needs of students
- Ensure that teachers and other relevant employees are aware of the importance of identifying children who have special educational needs

and "inculcate in students of the school an awareness of the needs of persons with disabilities" (Section 14, a-f).

Principals and SENO's must look to the future in planning for the education of children who will be continuing their education at another level, or beginning their training to enter adult life and the world of work. The Act requires them to ascertain the wishes of the child concerned and his or her parents. They must then take the necessary steps to enable the child to progress as a young adult to the level of education or training that meets his or her needs, the wishes of the parents, and what is appropriate to their ability.

For children who will be aged eighteen within the twelve months following a review of an educational plan, the NCSE will institute an assessment concerning the success of the goals in the previous plan, the reason for any failure to meet those goals and the effect this failure has had on the development of the child. This final plan must include measures to address any such effect.

Those services that fall under the remit of the Health Executive (e.g. speech and language therapy, physiotherapy, occupational therapy) must be performed and implemented according to the direction of the Minister for Education and Science and/or the Minster for Health. The Health Executive must appoint a liaison officer to coordinate and consult with the NCSE to ensure that all policies related to special education are consistent within both the education and health systems.

An important and influential aspect of the EPSEN Act is the establishment of the National Council for Special Education (page 188). The Council will eventually undertake total responsibility for special education services (it has already begun to undertake some).

There are many sections to this Act, too numerous to cover here. For the purpose of this book, I've concentrated on those of most relevance to parents, teachers and students preparing to become teachers. The Act itself is available to download from the Department Website (www.education.ie);

it is also available from the Government Publications Office, Molesworth Street, Dublin 2.

General Impression of the Education for Persons with Special Education Needs Act

The Act provides puts too much responsibility on school Principals who are already over-burdened with a huge number of tasks to complete. From assessment to the determination of the appropriateness of placement and planning, they have undue responsibility for the provision of special education. By eliminating the word 'Individual' from educational plans, the Act removes the child from the plan and makes it possible to create one-size-fits-all educational plans. The most significant flaw in the Act is its alarming ability to punish parents for pursuing court cases and its attempt to coerce them into mediation processes with people they may not necessarily trust.

The Act has its limitations but it is a major piece of legislation that will serve as the guiding template for all special education services and as such, I urge all readers to become very familiar with it.

Part Five

The Disabilities Act, 2005

This Act, the latest and perhaps what will be the last, piece of legislation designed to protect **the rights of the disabled**, was enacted into law in July 2005. It provides for the assessment of the health and educational needs of persons with disabilities and assures that appropriate planning will be undertaken on their behalf "…consistent with the resources available…" by the Ministers concerned. (Introductory Section)

The Act **defines disability** as "…a substantial restriction in the capacity of the person to carry on a profession, business, or occupation in the State or to participate in social or cultural life in the State by reason of an enduring physical, sensory, mental health or intellectual impairment." (Part 1, Section 2).

The term "substantial restriction" is defined as follows:
"A restriction which –
(a) is permanent or likely to be permanent, results in a significant difficulty in communication, learning or mobility or in significantly disordered cognitive processes, and
(b) gives rise to the need for services to be provided continually to the person whether or not a child or, if the person is a child, to the need for services to be provided early in life to ameliorate the disability" (Part 2, Section 7, 2).

The Act creates the position of **Assessment Officers** who have the responsibility for conducting, or arranging for the conducting of an assessment of a person with a disability. If the Officer concludes that there is a need for educational services, he or she must advise the National Council for Special Education who will see to it that appropriate educational services are provided. The Act states that Assessment Officers are independent in the performance of their duties.

The Act specifies the following about assessments:
• "An assessment under this section shall be carried out without regard to the cost of, or the capacity to provide, any service identified in the assessment as being appropriate to meet the needs of the applicant concerned" (Part2, Section 8, 5).

Assessment Officers, upon completion of the assessment are to provide a report that indicates "whether the applicant has a disability".
In cases where it is determined that the applicant has a disability, the report must include a statement on:
• The nature and extent of the disability
• The health and education needs (if any) occasioned to the person by the disability
• The services considered appropriate by the person or persons referred to in subsection 2 (the assessors) to meet the needs of the applicant and the period of time ideally required by the person or persons for the provision of those services and the order of such provision,

- The period within which a review of the assessment should be carried out" (Part 2, Section 8, 7).

In carrying out an assessment, the Assessment Officer "...may ... invite the applicant ... to meet with him or her for interview and furnish any documents or things relevant to the assessment in the possession of the applicant ... that he or she may reasonably request and the applicant shall comply with the request." (Part 2, Section 8, a). The Assessment Officer is to inform the applicant of the purpose of the interview "...unless in his or her opinion the provision of such information might be prejudicial to the applicant's mental health, well-being or emotional condition or inappropriate having regard to the age of the applicant or the nature of his or her disability." (Part 2, Section 8, b).

In the process of assessment, the Assessment Officer is to ensure that open communication and appropriate participation by the applicant is on-going. The applicant is to receive "...adequate information relating to the process of the assessment and the results of the assessment unless in his or her opinion the provision of such information might be prejudicial to the applicant's mental health, well-being or emotional condition or inappropriate having regard to the age of the applicant or the nature of his of her disability." (Part 2, Section 8, c, ii).

If it is a child that is being assessed, and the assessment indicates the need for educational services, the Assessment Officer is to contact the principal of the child's school (if enrolled in school), who will refer the matter to the NCSE "...for the purposes of assessment..." (Part 2, Section 9).

Who may apply for assessment?
Under the terms of the Act, any person who believes he or she may have a disability may apply to the Health Executive for assessment. If another person believes a potential applicant is not able to come to a reasonable conclusion about the need for assessment, the other person may make an application. This other person is defined in the Act as,
- "A spouse, a parent or a relative of a person referred to above

- A guardian of that person or a person acting *in loco parentis* to that person
- A legal representative of that person, or
- A personal advocate assigned by Comhairle to represent that person" (Part 2, Section 9).

A person who has made an application for assessment may make an additional application "...if he or she is of the opinion that since the date of the assessment-

- There has been a material change of circumstances
- Further information has become available which either relates to the personal circumstances of the applicant or to the services available to meet the needs of the applicant, or
- A material mistake of fact is identified in the assessment report" (Part 2, Section 9, 8).

The Service Statement

The Act stipulates that when an assessment report indicates that the provision of educational or health services (or both) are required, the Health Executive will "...arrange for the preparation by a liaison officer of a statement ... specifying the health services or education services or both which will be provided to the applicant by or on behalf of the Executive (Health Executive) or an education service provider, as appropriate, and the period of time within which such services will be provided" (Part 2, Section 11,2).

The NCSE will be contacted and requested to make such arrangements as are necessary to provide services. However, "The Council shall comply with a request ...unless it considers that-

- The assistance concerned is not required
- Such compliance would not be consistent with its functions or would unduly prejudice the performance of any of its functions, or
- Having regard to the resources available to it, it is not reasonable for it to comply with the request" (Part 2, Section 11, 3, c).

It is important to note that the Act restricts the writing of a service statement to adults or people outside of a school context. Part 7, Section

6 of the Act states, "A service statement shall not contain any provisions relating to education services where the subject of the statement is a child."

The Act makes it clear that an appeals process is in place and that it relates to both assessment and service statements. Since service statements will not relate to the educational planning of school children (except those matters which are the remit of the Health Executive such as speech and language therapy, occupational therapy, physiotherapy, etc.), we will look at the appeals process as it relates to assessment, with some reference to service statements.

The Complaint Process

A person may undertake a complaint about assessment by him or herself, or through a person referred to above in relation to any of the following:

- "A determination by the assessment officer concerned that he or she does not have a disability
- The fact, if it be the case, that the assessment ... was not commenced within the time specified ... or not completed without undue delay
- The fact, if it be the case, that the assessment ... was not conducted in a manner that conforms to the standards determined by a body (professional body which covers the services of the assessor)
- The contents of the service statement provided to the applicant
- The fact, if it be the case that the Executive (Health Executive) or the educational service provider, as the case may be, failed to provide or to fully provide a service specified in the service statement" (Part 2, Section 14, a-e).

Within ten days of receiving a complaint it will be forwarded to a Complaints Officer for review. Complaints officers are independent in their work. If the Officer determines that the complaint is "frivolous or vexatious" he or she may dismiss the complaint. A written report of the reasons for dismissal will be provided to the complainant, the liaison officer involved in the case, and sometimes the assessment officer concerned.

If the Complaints Officer views the complaint as legitimate, he or she will determine whether or not to initiate informal resolution procedures to address and resolve the complaint. Records of the resolutions arrived at will be kept and provided to the applicant and other officers as necessary. If a complaint is not resolved under the informal process, the Complaints Officer will keep a record of the matter and send a copy of it to the Health Executive who will then refer the matter to another Complaints Officer for investigation.

If the second Complaints Officer comes to the opinion that the matter is not suitable for informal resolution, he or she will contact all concerned, including the educational services provider if one is involved, and permit each to make a representation of their case. Following this representation of information from all concerned parties, the Complaints Officer will issue a written report of his or her findings to all concerned parties. The findings will include recommendations for further action.

The report of a Complaints Officer can include any or all of the following:
• That the case was, or was not, well founded
• That an assessment be provided and completed in a specified time period
• That further assessment take place if the person is found to have a disability
• That an assessment be carried out in conformity with standards of practice
• That the service statement be amended, varied, or added to by the liaison officer
• A recommendation that the service required in the service statement be provided

The Appeals Process
The Act specifies that there will be an Appeals Officer appointed who will be independent in the function of his or her duties. The Appeals Officer will submit an annual report to the Minister for Justice, Equality, and Law Reform. A person or persons acting for the applicant may make an appeal against a finding or recommendation of a Complaints Officer or against the Executive or educational service provider. Likewise, the Executive and educational service provider may make an appeal. Such appeals must be

filed within six weeks of the date from which the complaint results were submitted to the applicant. In certain conditions the Appeals Officer may extend the time period to twelve weeks. The Appeals Officer will make a written report of his or her findings to all concerned parties. The Appeals Officer may require parties to submit further information. The Appeals Officer has the power to require attendance at oral hearings and require the submission of all documentation related to the appeal. Failure to attend, submitting false information or obstruction of the appeals process is punishable by a fine of up to €3,000.

The Mediation Process
The Appeals Officer may authorise members of his or her staff to serve as "Mediation Officers". Any time after an appeal has been initiated, the Appeals Officer can refer the case to a mediation officer. If an applicant objects to this referral to mediation, the appeal will proceed as described above and the Appeals Officer will issue his or her written ruling. An appeal resolved by mediation will result in a written statement of the resolution and all concerned parties will sign it.

Part 2, Section 20, of the Act states, "An appeal to a court shall not lie against a determination of the Appeals Officer other than an appeal on a point of law to the High Court."

The remaining sections and parts of the Act relate to a wide number of disability issues including access to buildings, access to services, heritage sites, etc.

General Impression of the Disabilities Act
Like so much other legislation related to disability and special education needs, the Act contains the usual opt-out clauses, this time right from the introduction to the Act; 'Applications for assessment *may* be invited for interview prior to assessment. The applicant *must* reply to a request to submit information and documentation to the liaison officer involved in the case. The purpose of an interview does not necessarily have to be made clear to the applicant. Information about the process of assessment

does not have to be given to the applicant. These decisions all lie with the liaison officer. Service statements will contain recommendations which have 'regard to the resources available'.' Although flawed in these terms, the Act is a major step forward in providing legal rights and entitlements to persons with disabilities.

Departmental Guidelines Related to Special Education

The Department of Education and Science (DES) issues **'circulars'** from time to time, to the principal of every recognised school in the country. Circulars relate to every aspect of education and represent official Department policy; therefore they are important documents. They do not circumvent or replace legislation but are meant to add necessary detail concerning the day-to-day operation of schools in all their functions.

The most important policy concerning special education was issued in November 1998 by the then Minister, Mícheál Martin. This policy was the first to directly and cogently articulate a clear vision for the provision of special education in Irish schools. In this circular it was stated in unambiguous terms that all children with a special education need have an "automatic entitlement" to special education services and all the necessary supports and resources required to provide for their education. Additionally, the circular stated that the sort of resources and support required were to be determined *based upon the needs of the child* as an individual, rather than on the ability of the system to provide for those needs.

The following measures, all approved by the Government at the time, were included in this policy statement:
- The introduction of a formalised system of special teaching support on a fully integrated basis for all children attending schools who have been assessed as having special educational needs
- The introduction of a formalised system of child care support for all children with special needs, including those in special schools, special

classes and ordinary schools, who have been assessed as requiring such support
- Formal recognition of the distinct educational needs of all children with Autism, whose condition so requires the introduction of a special pupil teacher ratio of 6:1 for such children, together with an automatic entitlement to child care support

In issuing his policy statement Minister Martin stated, "For too long, the needs of many children with disabilities, particularly those in smaller groups or in isolated settings, have been supported in a reactive and entirely unsatisfactory manner. For too long parents have had to campaign tirelessly to give their children the chance to participate in and benefit from education." In the view of the author Minister Martin was a leading figure in championing the educational rights of children with special education needs and his role in setting in motion everything that has followed, both legislation and department policy, cannot be overstated.

Following the release of this important circular, a number of circulars have appeared which outline various duties and responsibilities as well as official Departmental policies relating to special education provision. Some of these indicate the process of determining how many specialist teachers are required in a given school and the ratio of teachers to pupils with a particular disability. It is not within the scope of this book to review every one of these circulars; the rapid pace of change related to these particular policies makes it likely that any such review would be outdated by the time of publication. However, two particular circulars deserve scrutiny, because they indicate clearly what the process of special education provision will be, from the time a special need is suspected to the time one is fully assessed and educational planning is undertaken.

Circular 08-99
This circular introduced the system of "Resource Teaching". It allocated differing amounts of time, in hours, that a pupil with an assessed special education condition would receive from a resource teacher. It also outlined the role of the resource teacher as follows:

A Resource Teacher assists schools in providing support for children with special educational needs arising from disability by:

(a) Assessing and recording child needs and progress;

(b) Setting specific, time-related targets for each child and agreeing these with the class teacher and principal;

(c) Direct teaching of the children, either in a separate room or within the mainstream class;

(d) Team-teaching – so long as the children concerned are deriving benefit from it;

(e) Advising class teachers in regard to adapting the curriculum, teaching strategies, suitable textbooks, use of Information Technology and suitable software and a range of other related matters;

(f) Meeting and advising parents, when necessary, accompanied by the class teacher, as necessary;

(g) Short meetings with other relevant professionals, in the children' interest – e.g. psychologists, speech and language therapists, visiting teachers, special school or special class teachers.

In addition to outlining the number of hours of resource teaching, and the role of the resource teacher, the circular also describes how an application for a resource teacher could be made by a school (the application is made by the Principal to the relevant Inspector):

The school must have the following information available to the Inspector when s/he calls:

- Name and date of birth of the child;
- Class in which the child is currently placed and level of attainment;
- Psychological reports and, where appropriate, other specialist reports e.g. audiological, speech and language therapy reports, etc;
- Confirmation that the school has received parental agreement regarding acceptance of support from a resource teacher in respect of each child;
- Confirmation that suitable accommodation is available or will be made available in the event of an appointment.

The Inspectorate will review the recommendations contained in the reports drawn up by the relevant professionals, as well as current provision available in the school and in the local area when considering applications. Applications in regard to children can only be considered when the relevant professional reports are made available.

The circular used the SERC report as the basis for establishing its guidelines for the allocation of resource teachers in schools and stated that in the future, the Department would work towards realising all the SERC recommendations in full, in terms of pupil-teacher ratios.

Circular 08-02

This circular rescinded circular 08-99 and made some adjustments to the pupil-teacher ratios as well as the process of applying for resource teaching hours. Like its predecessor, it also listed the criteria used to ascertain whether or not a pupil has a particular special education condition. The circular does not represent a significant enough change to warrant additional analysis at this time.

Circular Sp Ed 24-03

Although seeming to relate to the application and appointment process of special needs assistants, this circular, in its appendix, outlines what is called the 'staged approach to special education needs'. It is worth quoting the department's own words in this circular at some length because they relate to official policy about the general ethos of special education:

> "Although children with SEN may learn at a different pace and in a different way from other children, they need to belong to a peer group and to mix with children of different abilities in a variety of situations. Research on mixed ability teaching illustrates that children of lower ability benefit greatly and children of average or above ability are not academically disadvantaged. However, the practice has developed in recent years of using resource hours for individual tuition only. An exclusive reliance on this approach is contrary to the principle of integration in teaching and learning. Wherever possible,

Part Five

schools should provide additional help for children in the mainstream classroom or, if necessary, in small groups. This will also have the effect of minimising the disruption to the normal class programme that can happen if individual children are being withdrawn at different times for tuition." (p. 2,3)

The Staged Approach

This circular introduces what is called a "staged approach" to special education (as described in the prologue, page xi). There are three distinct stages in this approach, each designed to be sequential and helps to prevent a child from being placed prematurely in a special education setting, or from receiving special education services they might not need. Stage One is implemented when someone has reason to suspect the child might have a learning difficulty. In this Stage, the class teacher provides whatever minor assessment and intervention may be necessary. Stage Two commences when the interventions in Stage One have proven unsuccessful. In Stage Two, the Learning Support Teacher (see page 195) gets involved, but only with parental permission. Stage Three of the process is implemented only if the interventions in Stage Two have been unsuccessful. In this stage, more intensive assessment is undertaken to ascertain if there is a special education condition present. Like Stage Two, this stage is implemented only with parental permission.

Stage One

If either a parent or teacher has concerns about a child's physical, intellectual, academic, social, emotional or behavioural progress, the class teacher should administer screening measures (see glossary) to further assess the presence of, and nature of the difficulty, if one is present. Alternatively, the teacher might administer what is a 'norm-referenced rating scale' (see glossary) to get more detailed information.

The class teacher, in consultation with the parent, should then draw up a simple plan of intervention to take place within the classroom, to assist with the relevant areas of learning or behavioural difficulty. The plan should be reviewed regularly with the parent and implemented over a period of about two school terms. If concerns remain after two terms, the school's special education support team may be consulted about movement to 'Stage Two' of the special education process.

It should be noted that stage one of the process applies to children from Senior Infant classes and above. Children in Junior Infant classes are not covered under this process. (It is also important to note that the process outlined in this circular relates to children who have not yet been identified as having a special education need. All children who have been assessed and found to have a special need must have individual educational plans (see page 12) either in effect, or in the process of being drawn up).

Stage Two

At this stage of the process, the child should be referred to the Learning Support teacher (see page 195), with parental approval, for further diagnostic assessment. If this assessment indicates that supplemental teaching from the learning support teacher would be beneficial, then it can be provided (again with parental approval). The class teacher, the learning support teacher and the parent should be involved in the writing of an intervention plan, which should indicate how each will engage in supporting the child.

The learning support teacher, class teacher and parents should consult regularly. If concerns remain after about one school term, it may be necessary to move to stage three of the process.

In the case of children with significant behavioural difficulties, it may be necessary to consult the NEPS psychologist or a psychologist employed by the Health Executive. This can only be done with parental approval and may result in the drawing up of a detailed support plan for home and school, or may result in movement to stage three of the process. In any event, the process provides for more assertive intervention for those children who present with serious behavioural concerns.

Stage Three

At this stage, and if the interventions at stage two have proven unsuccessful, the school, with parental approval, may seek professional assessment from a relevant specialist. This may be an educational psychologist, clinical psychologist, speech and language therapist or medical professional. The class teacher, learning support teacher, resource teacher if necessary, parents, and outside professionals should then draw up a detailed learning plan which indicates all the supports necessary to implement the plan. This plan should be reviewed regularly with the parents and with specialists, and revised as necessary.

At the time of writing, this circular is still in effect although its implementation is far from sufficient. The reader is urged to question whether or not the process was implemented, how it was implemented (ask to see the written plan the class teacher developed), and who was involved in assessing its progress if the parents were not informed about it previously.

As is the case with all Departmental circulars, position papers and guidelines, including those that follow, parents and teachers must keep current with any changes occurring to them.

Parents and teachers should note that under the terms of Circular Sp Ed 02-05 (see page 178) any child receiving services under the terms and conditions of Stage Two or Three should proceed with the following two criteria in mind:

• "Intervention with pupils at stages II and III should include a classroom support plan to ensure that the pupils' needs are met for the whole of the school day.
• The development of literacy and numeracy skills will be a major component of many interventions at stages II and III. However, special educational needs in areas such as oral language, social interaction, behaviour and application to learning tasks may also need to be addressed."

Other important Department circulars

There are a number of other circulars that relate to special educational issues that require introduction at this time.

1. **Circular 32-03** relates to *grade retention* (keeping a child back to repeat a year in primary school). The circular makes the point that the curriculum is child-centred and can be adapted to meet a child's needs. It points out the level of support available to children with special needs in primary school that should enable children to move from grade to grade without retention. The circular states that children should only be allowed to repeat a year once, and then only for exceptional reasons and that a substantial number of children should not repeat a grade. If exceptional circumstances exist and the principal decides, in consultation with the class teacher and parents, that a child may benefit from repeating a grade, a record outlining the educational basis for this decision must be kept and a record of what new approaches will be implemented for the child must be included. Children in sixth class must not be changed from one school to another for the purpose of grade retention.

*I want it stated clearly and emphatically that I believe that children with special education needs do **not** benefit from grade retention. There are numerous citations in the special education literature making this point. Parents should resist any attempt by a teacher or principal to retain a child with special needs. The only appropriate interventions for children with special education needs are assessment and individual educational planning.*

2. **Circular M14-05** outlines the *grants available* for students with disabilities in secondary schools to purchase equipment. These grants are for children with serious physical and/or communicative disabilities that make ordinary spoken or written communication impossible. A grant of €3,800 is available to purchase equipment that is of educational value to them. Applications are processed through the SENO (Special Education Needs Organiser) assigned to the school.

Part Five

In the case of students with visual or hearing impairment, the visiting teacher service is also involved in processing the grant. A recent assessment is required to make the application and the assessor must have recommended that 'assistive technology' is necessary for the effective education of the child and must indicate how the equipment is to be used to support the child's education. It will have to be documented that the child needs the equipment throughout the school day and that the existing equipment in the school is insufficient to meet the child's needs. The equipment is the property of the school, for the use by the child while enrolled in the school. The equipment may be allocated to another school when no longer required, or if the child moves to another school.

3. **Circular Sp Ed 02-05**

   This circular is the most recent one to be released by the Department, appearing on its website on 4th September 2005. Information in the circular is consistent with previous circulars, it underscores the importance of circular 24-03 and it makes some minor changes in the content of some previous circulars. The circular relates to the organisation of resources for children with special education needs in primary schools.

   The circular outlines the key principles of what is called **The General Allocation Scheme**. These principals are:
   1. "Pupils needs can be met immediately
   2. Individual applications will continue for pupils with Low Incidence Special Education needs
   3. The level of support can be matched to the level of need
   4. One-to-one and group teaching are both possible
   5. Pupils should be supported by the most appropriate teacher"

In general these principles assure that all schools have sufficient resource teaching-hours for children with High Incidence special education needs (dyslexia, dyspraxia, etc.) and those children who require the services of a learning support teacher. Children with Low Incidence conditions (Autism, Asperger's etc.) will receive services on the basis of individual applications.

Schools will assure that those children with the greatest need get the most intensive level of support. Each school will decide for itself whether one-to-one or group teaching, or a mixture of both, is the best support for each individual pupil. The training and experience of a teacher will be considered when deciding which teacher assists which pupil.

The General Allocation Model
The General Allocation Model is designed to provide teaching resources to children who:
- Are eligible for Learning Support Teaching (below 10th percentile in reading or maths) (see page 245)
- Have learning difficulties, including pupils with mild speech and language difficulties
- Exhibit mild social or emotional difficulties
- Exhibit mild co-ordination or attention control difficulties (associated with dyspraxia or ADHD)
- Have needs arising from High Incidence disabilities such as Borderline General Learning Disabilities, Mild General Learning Disability and Specific Learning Disability (see pages 34 and 40)

The Circular goes on to state that most of the pupils having the conditions described above will receive support in the classroom, or in small groups out of the classroom or perhaps one-to-one teaching, for a period of time.

The circular makes a point of stating that the rationale of the General Allocation Model is to ensure we have 'inclusive' schools. It indicates that the terms 'Learning Support Teacher' and 'Resource Teacher' may change in the future, possibly to be referred to by a more general title such as 'Special Education Teacher'. However at the present time, (and in this circular), the titles remain unchanged.

Resource teachers can be assigned to individual pupils based on individual need or as part of the General Allocation Model. The circular further indicates, "all additional teaching support will build on, and complement, the support planned for, and delivered by the class teacher."

Part Five

An important section of the circular will be quoted in its entirely here.

"An essential principle of the general allocation is that the teaching resources made available under the model will be allocated to pupils according to their needs. Pupils with the highest level of need will, therefore have the highest level of support.

Principals and teachers will therefore need to have regard to the following considerations when allocating teaching responsibilities for pupils catered for by the general allocation, as well as for those pupils for whom resources have been allocated on the basis of low-incidence disabilities:

- It is the needs of all pupils who require additional support that should determine the manner in which full-time and part-time learning-support teachers and resource teachers are deployed.
- Whenever possible, pupils with the greatest need should be taught by teachers who have the relevant expertise and commitment and who have a degree of permanence of status that can guarantee continuity of provision.
- The training, experience and expertise of teachers should be taken into account by the principal when allocating teaching responsibilities in respect of pupils with learning needs at stages II and III.
- Logistical factors, such as timetabling for in-class additional teaching support and for withdrawal of pupils from mainstream classes, should be taken into account in order to ensure an inclusive approach to the education of the pupils to the greatest extent possible.

It is important that, where possible, schools should deploy experienced and qualified teachers to meet the needs of pupils with special educational needs."

The circular also adds some additional detail outlining how children come to be included under some of the special education categories recognised by the Department. Because this information is so important to the professionals who assess children and make a determination about the category of special need they have (if any), it will be quoted here in full:

"Physical disability

Such pupils have permanent or protracted disabilities arising from such conditions as congenital deformities, spina bifida, dyspraxia, muscular dystrophy, cerebral palsy, brittle bones, or severe accidental injury. Because of the impairment of their physical function they require special additional intervention and support if they are to have available to them a level and quality of education appropriate to their needs and abilities.

Many require the use of a wheelchair, mobility or seating aid, or other technological support.

They may suffer from a lack of muscular control and co-ordination and may have difficulties in communication, particularly in oral articulation, as for example severe dyspraxia.

Pupils with a physical disability who have learning difficulties arising from the disability may need resource teaching where there are consequent significant learning difficulties. Others may need assistive technology only.

Hearing impairment

Such pupils have a hearing disability that is so serious to impair significantly their capacity to hear and understand human speech, thus preventing them from participating fully in classroom interaction and from benefiting adequately from school instruction. The great majority of them have been prescribed hearing aids and are availing of the services of a Visiting Teacher. *(This category is not intended to include pupils with mild hearing loss.)*

Schools that have a pupil who has been assessed as having hearing impairment and no other assessed disability, may be allocated a maximum of 4 hours teaching support per week from a resource teacher, or from a visiting teacher and resource teacher combined.

Part Five

Where a pupil with a hearing impairment also meets the criterion for another low-incidence disability category, provision is allocated as for multiple disabilities.

Visual impairment

Such pupils have a visual disability which is so serious as to impair significantly their capacity to see, thus interfering with their capacity to perceive visually presented materials, such as pictures, diagrams, and the written word. Some will have been diagnosed as suffering from such conditions, such as congenital blindness, cataracts, albinism and retinitis pigmentosa. Most require the use of low-vision aids and are availing of the services of a Visiting Teacher. *(This category is not intended to include those pupils whose visual difficulties are satisfactorily corrected by the wearing of spectacles and/or contact lenses.)*

Schools that have a pupil who has been assessed as having a visual impairment, and no other assessed disability, may be allocated a maximum of 3.5 hours teaching support per week from a resource teacher, or from a visiting teacher and resource teacher combined.

Where a pupil with a visual impairment also meets the criterion for another low-incidence disability category, provision is allocated as for multiple disabilities.

Emotional disturbance and/or behaviour problems

Such pupils *are* being treated by a psychiatrist or psychologist for such conditions as neurosis, childhood psychosis, hyperactivity, attention deficit disorder, attention deficit hyperactivity disorder, and conduct disorders that are significantly impairing their socialisation and/or learning in school. *(This category is not intended to include pupils whose conduct or behavioural difficulties can be dealt with in accordance with agreed procedures on discipline.)*

Some pupils in this category *may* need resource teaching support. Care support from a special needs assistant may be required where a pupil's behaviour is a danger to himself or others or where it seriously interferes with the learning opportunities of other pupils. In certain circumstances, some pupils may require both supports.

Moderate general learning disability
Such pupils have been assessed by a psychologist as having a moderate general learning disability.

A maximum allocation of 3.5 hours teaching support per week from a resource teacher may be made to schools in respect of each pupil assessed as having a moderate general learning disability (the pupils full-scale IQ score will have been assessed in the range 35-49).

Severe or profound general learning disability
Such pupils have been assessed by a psychologist as having a severe or profound general learning disability. In addition, such pupils may have physical disabilities.

Five hours teaching support per week from a resource teacher may be made to schools in respect of each pupil with a severe/profound general learning disability (the pupil's full-scale IQ score will have been assessed as being below 35).

Autism/autistic spectrum disorder (ASD)
A psychiatrist or psychologist will have assessed and classified such pupils as having autism or autistic spectrum disorder according to DSM–IV or ICD–10 criteria.

In the interest of the pupil with an ASD and in order that the needs of the pupil are adequately addressed, it is important, where feasible, that for a definitive assessment of ASD, a multi-disciplinary assessment team should be involved. The need for a multi-disciplinary assessment is also in keeping with the policy of the National Educational Psychological Service (NEPS).

Part Five

A maximum allocation of 5 hours teaching support per week from a resource teacher may be made to schools in respect of each pupil assessed as having ASD.

Pupils with special educational needs arising from an assessed syndrome
The level of additional support to be provided for pupils who present with a particular syndrome e.g. Down syndrome, William's syndrome and Tourette's syndrome will be determined following consideration of psychological or other specialist reports which describes the nature and degree of the pupils special educational needs.

Where a pupil with an assessed syndrome has a general learning disability, resource teaching support will be allocated to schools in line with hours allocated to pupils assessed as being within the same IQ band (moderate/severe/profound GLD). Where a pupil with an assessed syndrome has any of the other low-incidence disabilities, resource teaching support will be allocated on that basis.

Specific speech and language disorder
Such pupils should meet each of the following criteria:
- The pupil has been assessed by a psychologist on a standardised test of intelligence that places non verbal or performance ability within the average range or above.
- The pupil has been assessed by a speech therapist on a standardised test of language development that places performance in one or more of the main areas of speech and language development at two standard deviations or more below the mean, or at a generally equivalent level.
- The pupil's difficulties are not attributable to hearing impairment; where the pupil is affected to some degree by hearing impairment, the hearing threshold for the speech-related frequencies should be 40Db;
- Emotional and behavioural disorders or a physical disability are not considered to be primary causes.
- Pupils with speech and language delays and difficulties are not to be considered under this category.

- In the case of specific speech and language disorder it is a pupil's non-verbal or performance ability that must be within the average range or above. (i.e. non-verbal or performance IQ of 90, or above).
- The pupil must also have been assessed by a speech and language therapist and found to be at two or more standard deviations (S.D.) below the mean, or at a generally equivalent level (i.e. – 2 S.D. or below, at or below a standard score of 70) in one or more of the main areas of speech and language development.
- Two assessments, a psychological assessment and a speech and language assessment are necessary in this case.
- A maximum allocation of 4 hours teaching support per week from a resource teacher may be made to schools in respect of each pupil assessed as having specific speech and language disorder.

Multiple disabilities

Pupils assessed with multiple disabilities meet the criteria for two or more of the disabilities described above. A maximum allocation of five hours teaching support per week from a resource teacher may be made to schools in respect of each pupil assessed as having multiple disabilities.

Applications for resources for pupils with special educational needs arising from low-incidence disabilities should be made to the assigned Special Educational Needs Organiser."

How the General Allocation Model Works

The General Allocation Model, according to the Circular, has advantages over the previous system which was proposed in June 2004. It provides for the allocation of permanent teachers, based on enrolment, to provide educational services to children with both learning difficulties and special education needs. The circular suggests its advantages as:

- "It provides schools with permanent resources for pupils with special educational needs arising from high incidence disabilities and thereby facilitates flexible and early intervention for these pupils;
- It reduces the need for individual applications and supporting psychological assessments for pupils with special educational needs arising from high incidence disabilities;

Part Five

- It provides resources more systematically, thereby giving schools more certainty about their resource levels;
- It gives greater flexibility to school management in the deployment of resources.

It provides greater levels of certainty about resource allocations thereby facilitating better planning within the system both at central and local level, leading to a more effective and efficient delivery of services."

The General Allocation Model works differently in large schools and small schools. In larger schools the formula is as follows:
"Differing pupil teacher ratios apply to boys', mixed and girls' schools.

- Boys' schools with 135 pupils or more get their first post at 135; second post at 295; third post at 475, fourth post at 655, and so on.
- Mixed schools with 145 pupils or more get their first post at 145; second post at 315; third post at 495, fourth post at 675, and so on.
- Girls' schools with 195 pupils or more get their first post at 195; second post at 395; third post at 595; fourth post at 795, and so on.
- All designated disadvantaged schools get their first post at 80; second post at 160; third post at 240; fourth post at 320, and so on.

It should be noted that schools qualify for a pro rata part of a post for pupil numbers below the enrolment point for the first post and between the first and second post, the second and third post, and so on. For a designated disadvantaged school with 60 eligible pupils the general allocation is 0.8 of a post; for a boys' school with 215 pupils the general allocation is 1.5 posts; for a mixed school with 700 pupils the general allocation is 4.1 (rounded to one decimal place)."
In small schools the General Allocation Model formula is:
  Boys' small schools (less than 135 children) get their first post at 100 children
  Mixed small schools (less than 145 children) get their first post at 105 children
  Small girls' schools (less than 195 children) get their first post at 150 children

No additional allocation is going to be made to small boys' schools between 100 and 135 children; to mixed small schools between 105 and 145 children, or to girls' small schools between 150 and 195 children.

General Impressions of Circular 02-05

This is one of the most important circulars the Department has issued in recent times. It outlines the model upon which supplementary teachers will be assigned to schools. It underscores the importance of Circular 24-03 and places renewed emphasis on the important role of the classroom teacher in being the person with primary responsibility for the education of all children. It provides a formula upon which schools will automatically receive supplementary teaching posts, thereby guaranteeing that every primary school in the country will have specialist teachers.

The major flaw in the circular is its reliance on the total number of children enrolled in the school to determine the number of specialist teachers. Although children with Low Incidence disabilities will continue to have teaching resources provided to them on a case by case basis, and through an application process, it is likely that the number of children with High Incidence disabilities is going to vary quite a bit from one school to another. Therefore, a strict adherence to a formula in staffing ratio may present problems to schools in which many children have been identified with these High Incidence conditions.

Once again I want to underscore the critical importance of regularly accessing the DES and SESS websites for circulars related to special education (see list of useful Web Sites, Page 263).

Structures in Irish Special Education

Only ten years ago there were essentially no structures in Irish special education. A large number of segregated special education schools operated as a system outside the mainstream, offering at times excellent services to children with special needs. The international emphasis on integration and inclusion has compelled the DES to create structures in response to these new developments in special education. What has been created is a vertical system i.e. one that operates from a national level down to the local (school) level.

The National Council for Special Education (NCSE)

Originally this Council was created after the Education Act, 1998 was passed. Upon ratification of the Education for Persons with Special Education Needs Act, 2004, the original body was disbanded (it had never really assumed any influential position) and this new body was created. The NCSE has the following functions under this Act:

- Dissemination to schools, parents and others of appropriate information about best practice in special education
- Planning and coordinating the provision of education and support services to children with special education needs
- Planning for the integration of children with special needs with their "non-disabled" (the term used by the Department) peers
- Providing information to parents about their entitlements and the entitlements of their children with special needs
- Ensuring that the progress of children with special needs is monitored and regularly reviewed
- Assessing and reviewing the resources required to provide special education services
- Ensuring that a continuum of provision is available as may be required by any type of disability (This means that depending on the type and nature of the disability, differing levels of support and differing intensity of support may be provided)
- Reviewing in a general sense the provision for adults with disabilities to avail of higher education
- To advise educational institutions about best practice related to adults with disabilities
- To advise the Minister about matters relating to education of children and others with disabilities
- Consultation with voluntary bodies as it considers appropriate
- Conducting and commissioning research on matters related to its work

As can be seen from the list of duties, the Council has nearly exhaustive responsibilities for the provision of special education services for the citizens and residents of Ireland. The Council is composed of a chairperson and twelve members. Gender balance is represented on the Council with six male and six female members and the membership is representative of educationalists, trade unionists, parents, teachers, members of the National Disability Authority and a member appointed by the Minister for Health and Children.

An **Appeals Board** will be established shortly as well. The Appeals Board will have responsibility to hear appeals made under the Education for Persons with Special Education Needs Act and to see to it that the timeline for hearing and deciding appeals is adhered to. The Appeals Board is independent in its functions.

One of the first tasks of the Council was to create a post of **Special Education Needs Organiser (SENO)** and to begin the hiring of people to assume this post. At this time about seventy SENO's have been hired and are currently working. The role of the SENO is to coordinate and advise schools at the local level in the provision of special education services. A SENO may be assigned to a particular area of the State and oversee a particular number of schools in that area, or may be assigned to a particular category of school. The staff members of the schools they service are obliged to cooperate with them in the performance of their duties. SENO's are supposed to have a qualification, expertise and experience relevant to the education of children with special needs.

SENO's work with schools and in particular, they work with the special education support team within a particular school. For a current list of contact information about SENO's that are assigned to each region in Ireland, access the following website: http://www.ncse.ie/docs/SENO_ office_addresses.pdf. Every school in Ireland should have a designated special education support team. Its exact composition is up to the principal but typically a learning support or resource teacher will be a member of the team along with the principal or his or her designee.

Part Five

Classroom teachers may be members of the support team as well. The support team works as a cooperative group for the benefit of those children in the school with special needs. Their role is to communicate, advise, collaborate and review special education provision for the children within the school and to work in close cooperation with the SENO.

On 14 July 2005, the Minister for Education and Science, Mary Hanafin announced that the NCSE will be formally established on 1 October 2005. In addition to the duties outlined above, the NCSE will also review the provision made for adults with disabilities to enable them to avail of higher, adult and continuing education. In her announcement the Minister also ordered that the Special Education Appeals Board be established on 3rd April 2006. Finally, the Minister ordered that thirty six sections of the Education for Persons with Special Education Needs Act be implemented. One of the more important sections of the Act to be implemented by the Minister will provide statutory force to inclusive education. However, the reader will recall that this statutory entitlement to inclusive education is based on the assumption that it will be in the best interests of effective provision for other children in the mainstream environment.

The National Educational Psychological Service (NEPS)

Established in September 1999 as a division of the Department of Education and Science, NEPS is the agency responsible for all educational psychological services to children (clinical psychological services are provided by the Health Executive). According to the documentation provided by NEPS, a psychologist will:

- Engage in individual casework with students with special needs
- Collaborate with others in evaluating appropriate interventions
- Consult with teachers and parents
- Contribute to 'whole-school planning' related to special education
- Contribute to the professional development of teachers

At the time of writing there are about one hundred and fifty psychologists working in the NEPS service. Every school in the country is supposed to have access to a NEPS psychologist, but in some regions this is not the case. As new psychologists are hired, every school will have a psychologist assigned to it but each psychologist may be responsible for a large number of schools or children, making it difficult for them to consult and visit schools on a frequent basis. The NEPS service is free to parents – there is no cost involved in having a child assessed psychologically through NEPS. It may be the case that a full, formal psychological assessment is not required and the NEPS psychologist can provide a valuable consultation to teachers and parents around issues of mild learning and behavioural or adjustment problems.

What is the appropriate role for an educational psychologist?
Ideally an educational psychologist provides a lot more than assessment services. The educational psychologist is an expert in how children learn; this knowledge, when combined with the expertise of the teacher, can be a great asset to teachers and school principals in deciding how best to help children learn. He or she will, (when the service is operating at a mature level) be available to mainstream teachers well before a child is suspected of having a special education condition. The educational psychologist can provide valuable input into curriculum differentiation (see page 116), educational strategies, methods of discipline, management of groups, and other essential information about how to help children to grow, develop, learn and achieve.

A NEPS psychologist has responsibility for the well being of the whole child and they must avoid becoming overly concerned about *academic* progress. Unfortunately, the reality of the caseload of the NEPS psychologist affords little opportunity for taking a 'holistic' view and it is difficult for them to achieve continuity of service provision. Having responsibility for so many schools, the role is often relegated to psychological 'gatekeeper'. That is, the psychologist is the person who provides the essential information making it possible to determine whether or not the child is eligible for special education services and what form these services will take. This restricted role places severe limitations on the benefit that can be accrued from access to an educational psychologist.

Part Five

Until the NEPS service is adequately staffed and resourced, with all the essential assessment tools and resources it will function mostly as a gatekeeper service.

The State Examinations Commission (SEC)

This body is responsible for the Junior Certificate and Leaving Certificate examinations. The SEC facilitates applications for 'reasonable accommodations' for children with special education conditions who are sitting these exams. Students with 'permanent or long-term conditions' are eligible to apply for these accommodations when taking the Leaving Certificate Exam. Accommodations are intended to remove the impact of the disability on the child's performance and to prevent undue advantage falling on the child who is accommodated. There is a range of accommodations that can be made on State examinations.

If a candidate's condition makes it impossible for him or her to participate in a particular mode of assessment, an accommodation can be made and it will be indicated on the exam results that an accommodation was made (no mention of the special condition will be recorded). If an accommodation is granted, the exam results will be calculated on the parts of the exam actually taken.

On written examinations the range of accommodations includes:
- Papers may be read to the student and questions repeated as often as required
- Papers may be modified, with questions that require the reading of charts and diagrams substituted for others that don't require this
- Braille translation of questions may be provided
- Papers may be provided in large print
- Candidates may record their answers on typewriters, word-processors or tape recorders
- Drafting machines, drawing boards and smaller drawing sheets may be used for some technical exams
- Answers may be dictated to a scribe rather than be tape-recorded
- Ten minutes extra answer time for each hour of examination allotted may be given

- If exams are to be recorded, the candidate may sit in another room, or the candidate can listen to the questions with headphones and sit in the main exam centre

Schools may permit the following accommodations without requesting permission from the Exam Commission:
- Breaks and rest periods may be allowed and a maximum of twenty minutes extra exam time may be allowed if breaks are taken
- Food, drinks or medication may be taken into the exam centre
- The candidate may be allowed to move within the exam centre
- A special desk or chair may be provided
- Low vision aids normally used in the classroom may be permitted
- Candidates with hearing impairment must be seated close to the exam superintendent

Applications for accommodation must be made directly to the Exams Commission. For cases in which accommodations are denied, there is an independent appeals process in place.

For students sitting the Junior Certificate Examination, similar accommodations can be made and applications and appeals are processed by the Exam Commission as outlined above.

Particular mention is made, regarding accommodations on State examinations, to Specific Learning Disabilities (dyslexia, dyscalculia, dysgraphia, dyspraxia). The condition must contribute to "marked failure to achieve expected levels of attainment in basic skills such as reading and writing". In such cases, reading assistance can be provided but only if the candidate's inability to read the exam questions arises directly from the specific learning disability. This means the candidate must have a severe reading disability. For language subjects, the candidate will obviously not be able to read the questions; therefore the results will be accompanied by an explanatory note that reads "All parts of the examination in this subject were assessed except the reading element".

Part Five

A tape recorder or computer may be used if it can be demonstrated that the candidate has good oral ability and good knowledge of course content and scores below average on a spelling test, with no more than 20% of the target words unrecognisable. The candidate's difficulties must always be a *result* of a specific learning disability. It is always the case that a recent comprehensive assessment of learning profile and related academic skills is necessary.

A helper can be assigned to assist with the housekeeping duties related to the Home Economics examination but may not provide factual assistance.

The reader is encouraged to keep careful track of these accommodations on the exams commission website (www.examinations.ie).

The School Staff Involved in Special Education
The Role of the Principal
Anyone who has read the Education of Persons with Special Education Needs Act can easily see the vital and central role played by the school principal. From beginning to end, the principal is directly involved; even if he or she designates a teacher to represent them in the process of providing special education services, the principal retains the ultimate responsibility for their implementation.

The Class Teacher
Department policy clearly indicates that regardless of what special education need a child may have, or who may be involved in supporting that need, the classroom teacher retains the primary responsibility for the education of all children in his or her classroom. There has always been a subtle temptation to turn responsibility for the teaching of reading and mathematics to children with special education needs over to a specialist teacher. This has resulted in a tendency for classroom teachers not to teach this content to these children. The Department is clearly indicating that the class teacher will always have responsibility for teaching *all* the curricular content to *all* the children and the classroom teacher has both general and specific responsibilities, particularly as outlined under Circular Sp Ed 24-03.

Supplementary Teachers

There are several groups of qualified teachers working in the system who share responsibility, along with the classroom teacher, for the education of children with special needs. These include:

- The Learning Support teacher
- The Special Education Resource teacher
- Visiting teachers (for children with visual and hearing impairment).
- The Language Support teacher
- The Support Teacher

The Department is moving towards referring to all of these teachers as 'Special Education Teachers', so the individual categories are likely to disappear in the future.

The Learning Support Teacher

Previously known as 'Remedial Teachers', the learning support teacher plays a major role: assisting children with academic weakness who do not have special education needs and in providing educational support to some children with special education needs, particularly those children with Dyslexia and Borderline Mild General Learning Disability. There are at present count about fifteen hundred learning support teachers in the system; every primary school and many secondary schools have one assigned to them, at least on a part-time basis.

Learning support teachers have an important role to play in the process of identification of children who may have special education needs but who have not yet been assessed. It is the role of the learning support teacher to work in cooperation with the classroom teachers to consider those children who are the weakest in literacy or mathematics in the classroom. The learning support teacher, based on the severity of their need, may then give these children 'small group tuition'. The learning support teacher and the classroom teacher are to work together in developing an intervention plan that will outline strategies of assistance to be provided by each of them. This plan will also be developed in cooperation with parents in so far as is possible to do so. The support

Part Five

provided by the learning support teacher typically extends over one school term (about thirteen to twenty weeks) and is then reviewed. The results of this review will determine if another term of supplementary teaching is required (or whether an assessment is required).

Successful outcome in learning support teaching depends on a number of factors including a supportive home environment. However, the most significant factor is the collaborative relationship between the learning support teacher and the classroom teacher. Ideally, learning support teaching should be a combination of tuition in the learning support teacher's room and tuition in the child's mainstream classroom. Learning support services are to be provided to the lowest achievement children in the system and to those with some of the more 'high incidence' disabilities such as dyslexia or borderline mild general learning disability. The time allotted to children receiving learning support services can vary significantly from school to school but typically a child may receive two or three thirty-minute sessions each week.

As is the case with all specialist teachers within the system, learning support teachers may or may not have education beyond the basic qualification. There are several higher diploma programmes available for learning support teachers, some provided by the Colleges of Education and some provided by the University departments of education. The quality of learning support service depends in no small measure on the extent of post-graduate education obtained by the teacher him or herself.

The role of the learning support teacher is:
- Collaboration and consultation at the whole-school level
- Collaboration with the Principal
- Collaboration with class teachers
- Collaboration with parents
- Coordinating selection of pupils for supplementary teaching
- Provision of supplementary teaching to individual and groups of children
- Conducting assessments and maintaining children's records

Educational Planning by Learning Support Teachers

Learning support teachers are urged in their guidelines to complete a 'Weekly Planning and Progress Record' for each student they teach. On this form they will write:

- The curricular area being addressed that week
- An overall programme of activities to be implemented that week
- Their observations of the child's learning and their assessment of that learning

This information is to be recorded weekly by the learning support teacher in consultation and cooperation with the class teacher and the parents. This sort of regular record keeping contributes to a portfolio of the child's work and progress over many months and can be a useful source of information for additional programme planning and review. The guidelines recommend that at the end of every term, usually thirteen to twenty weeks, the child's programme should have an in-depth review. Based on this review, decisions will be made about the sort of support the child requires in the future or whether he or she continues to require support at all.

It is informative to quote from the guidelines themselves about the issue of where learning support services should take place:

> "Supplementary teaching will sometimes take place in the learning support teachers' room rather than in the regular classroom. However when withdrawal for supplementary teaching is complemented by the implementation of collaborative approaches and direct intervention in the classroom then real and sustained progress can be made by pupils and the negative aspects of regular withdrawal from class can be minimised." (p. 77).

Parents should take careful note of this statement, for it contains an important message: if the learning support teacher relies only on withdrawal support, removing the child regularly from their mainstream classroom, real and sustained progress is unlikely to be made and there

Part Five

are negative aspects to this practice. Classroom teachers need to realise *their* critical role in providing supplementary teaching and supporting children with special needs. Parents need to inquire where the child is receiving support and what sort of collaboration exists between the learning support teacher and the class teacher.

The Special Education Resource Teacher

There are currently two thousand six hundred special education resource teachers in Ireland. In 1998 there were one hundred and four! This huge increase in number signals a major change in the provision of support services for children with special education needs. The vast majority of these teachers are employed in primary schools but increasing numbers are being allocated to secondary schools. The role of the special education resource teacher is confined to providing services to those children who have been assessed and found to have a special education condition.

Because the position of resource teacher is a relatively new one, it is not as clearly defined by policy documents as the learning support teacher. For the time being, the resource teacher functions in a grey area and the result is that there are different support structures in different schools.

The resource teacher is providing educational support to those children who have special education conditions as defined by the DES (see page 21). Therefore, they are the neediest children in the system, needy in the sense that they require the most intensive support. The major difference between the role of the resource teacher and the learning support teacher is one of time. Typically, children receiving services from a resource teacher receive more time per week than those receiving services from a learning support teacher. It is not uncommon for a child to receive up to two hours a day of resource support as opposed to the two to three half-hour sessions allocated to children receiving learning support services.

Like the learning support teacher, the resource teacher should be a qualified teacher and will therefore be considered a fully qualified special education teacher by having completed the basic credential programme, which is the B.Ed. or H.Dip (Primary) training programme. At present many of them do not have post-graduate qualifications in the area of special education, but the Colleges of Education are creating diploma courses for resource teachers and many are availing of this advanced educational opportunity.

The Visiting Teacher Service

The visiting teacher service provides educational support to children who are visually or hearing impaired and enrolled in mainstream schools. These teachers may provide some direct instruction but a significant aspect of their role is their ability to serve as consultant to the classroom teacher in adapting teaching methods and in using assistive technology in the classroom. Typically a visiting teacher will serve a large number of schools, as needs dictate.

The Language Support Teacher

This is a relatively new post, created in the last year or two in response to the increasing numbers of children enrolled in Irish schools who do not have English as their first language. The language support teacher will work with these children to develop language skills and make the transition to the education system here (and the transition to society) easier for them. As of yet there are no clear guidelines for the provision of this service.

The Support Teacher

This group of specialist teachers was established in the late 1990's, in recognition that a significant number of children were experiencing behavioural disturbances severe enough to interfere with their learning and the learning of their peers. The Support Teacher was initially designated as the Teacher Counsellor but the Department of Education and Science changed the name to Support Teacher in 1998. There are currently about forty-five Support Teachers in Ireland, the majority of them located in the greater Dublin area.

Part Five

The role of the Support Teacher, according to the departmental guidelines, is to provide help to children who are "disruptive, disturbed or withdrawn" to manage their behaviours and to enable the schools to teach these children and their peers effectively. The Support Teacher works on a referral basis with the neediest children in the school, working with them in small groups, in consultation and collaboration with family and class teachers. A variety of strategies are used to assist these children in the development of self-control.

As stated, all of the above posts in special education are supposed to be occupied by qualified teachers. However, it is a reality that due to the shortage of teachers in Ireland, some, who are not fully qualified, assume these posts. This is a rather contentious issue in Irish education at present and one that extends beyond special education. The Irish National Teachers Organisation (INTO) and the DES are working together to resolve this issue and progress is being made.

There is another group of people working in schools who do not have to be fully qualified teachers in order to assume their position. These are the special needs assistants.

The Special Needs Assistant (SNA)

The special needs assistant does not necessarily hold an educational qualification, (although some do have advanced degrees and educational qualifications). The entry requirement for this position is a Pass in Irish, English, and Mathematics at Junior Certificate level. The special needs assistant is an important member of the support staff for children with special needs. Six years ago there were six hundred SNA's in the system – at time of publication there are six thousand! They function in a clearly defined role and work under the direct supervision of the class teacher at all times. They are not to assume a direct teaching responsibility.

Although there is no requirement for a SNA to be trained to assume the position, many have earned certificates or diplomas. The DES has a training programme in place and at least one of the Colleges of Education offers both certificate and diploma level training opportunities for SNA's who wish to avail of them. The primary duty of a SNA is to support the child in non-academic tasks in the classroom such as:

- Mobility for those with physical disabilities
- Assisting with organisation and task completion, for those who might be unusually distractible
- Providing guidance during free play situations for those with behavioural problems
- Assisting children with autistic spectrum disorders in managing the school day in a mainstream or special setting

SNA's are not granted automatically to children with special education needs. The Department has recently turned over the application process to the National Council for Special Education. Following assessment, a school may submit an application for a SNA; the success of the application will be determined based on the severity of the child's need.

Conclusion

The careful reader might have become aware at this point of the many contradictions between different pieces of legislation, policy, and structures. One law gives parents the right to choose a school for their child; another allows an agency to dictate the school a child must attend. One law states that parents are to be involved in every phase of their child's special education; another states that a principal may take it upon himself or herself to make decisions and subsequently inform the parents of that decision. Educational planning is to be undertaken in cooperation and consultation with parents in one policy document and in another, teachers can work together to make plans, implement them, and meet with parents 'in so far as is possible'. Appeals procedures are established in at least three separate pieces of legislation but the appeals process within each is not yet fully established. Parents may appeal to a Council, to the Minister, to an appeals board, to a school principal, to the Inspectorate, etc. There seems to be little cohesion in the various structures, as they currently exist.

Part Five

Although legislation has been passed, much of it **lacks the necessary specific language or imperatives**. For example, the law specifies that Educational Plans are to be written but does not use the terminology 'individual' educational plans, a phrase that has international recognition as being at the core of all special education services. This means that the 'I' in the educational plan, the child him and herself, is missing from the most important piece of special education legislation existing in Ireland. Additionally, while the law requires that all resources and supportive services are to be provided, and specifies that the National Council for Special Education *can insist* that the Health Executive provide those services, the phraseology of the legislation is fraught with words such as "may", "as resources permit" and "in so far as possible". It is true that legislation cannot create human beings with specific skills necessary to support children with special needs, such as speech and language therapists; however, legislation without teeth is bound to be impotent.

The law states that children are to be assessed if there is a suspected special education need, and educational interventions in the first or second phase have not been successful. The legislation follows on to include principals and teachers in the list of those who are qualified to complete these assessments. Assessment is a special skill requiring post-graduate qualification in speciality areas. No teacher, principal, specialist teacher or other educator who has not undertaken such post-graduate education can make a reliable assessment of the presence or absence of a special education condition. It is a sad reality that in Ireland, teachers who were in education more than ten years ago received almost no input about special education in their degree programmes. Even today, the amount of input at pre-service level is barely adequate to cover a general review course. It is unreasonable to expect people who have not had access to advanced education to make meaningful assessments about special needs conditions.

Appeals processes are being put into place but the nature of how they will play out in reality remains unknown. There are also multiple paths to appeal: The Teaching Council, the National Council for Special Education, the Minister or the school principal. For a parent, the navigation through this maze of appeals will be a major undertaking and while appeals are being heard there is nothing in the legislation or policy documents that states specifically what will be done for the child in the meantime. Of course, a parent still has recourse to appeal through the courts as so many have done in the past ten years (see page 207). However, the legislation actively puts obstacles in the way of the parents' right to appeal to the courts, by not only requiring mediation, but by specifying that the results of mediation, including statements about the parent's refusal or good faith, may be entered into court proceedings to assist in determination of the awarding of costs.

An additional obstacle to the provision of appropriate special education services is the critical **shortage of speech and language therapists**, occupational therapists, physiotherapists, and psychologists in Ireland. This makes it increasingly difficult for children with special education needs to receive what they most desperately require: early intervention. Unfortunately, there is no easy solution to this problem.

The critical issue of just who is a 'qualified' special education teacher must be addressed ultimately. The notion that *any* graduate of a college of education or a Montessori training programme is also a fully **qualified special education teacher** is inappropriate (Montessori methodology is not necessarily sufficient for all children with special education needs). For a graduate of a B.Ed. or H.Dip Primary programme to be able to leave college in May and take a position in a school for children with severe autism in September is grossly inappropriate.

While it is true that the good heart of the teacher in special education may be their most important asset, it is foolish to disregard the need to acquire advanced skills in diagnostic assessment, educational planning and intervention; they must know to critically review goals and objectives, and have the specific skills necessary to teach children with specific disabilities. Society does not allow college graduates with university chemistry degrees to perform surgery. While the stakes may not be so

Part Five

high in special education, it is nonetheless the lives of children that are being shaped forever.

Before leaving this section on the legislative, directive, and structural framework of special education provision, I want to outline what works best when all these complicated and intricate systems and networks are put into place.

What Works Best?

What works best is continuous *cooperation* by all involved. Though access to some services may be fragmented, and in some cases, quite difficult indeed, communication is the best asset in special education. Parents know the child best, teachers know best how to teach the child, specialist teachers know best how to teach to particular special education conditions, educational psychologists know best how a particular child learns and behaves. Other specialists have knowledge about particular areas of children's functioning. Sharing all this important information in an atmosphere of genuine equality will ensure the child gets the most appropriate special education services. It is imperative for all concerned to put anger and distrust aside and work together for the child.

PART SIX

Special Education and the Courts

Over the past decade, a number of cases have been brought to the High Court by parents which relate to the education of children with special needs. It is largely as a result of these cases that the improvements we see in educational provision for such children have arisen. It is an unfortunate reality that wherever special education services are provided, the courts have been in a position of influencing their development, either directly or indirectly. There are two High Court cases that have special significance for the education of children with special needs. Although the scope of this book does not permit an exhaustive scrutiny of their development, it might be helpful to review them, with a view to gaining insight into their eventual impact on the special education system in Ireland.

Note: Information on Irish High Court Cases can be accessed at the Irish Legal Information Initiative on the Internet: www.ucc.ie/law/irlii/irliiindex/lc.php

Before doing so however it is essential to make note of the Irish constitution and what it has to say about the education of children.

The Irish Constitution
Article 42 of the Irish constitution deals with education. The full text of Article 42 is as follows:

The State acknowledges that the primary and natural educator of the child is the Family and guarantees to respect the inalienable right and duty of parents to provide, according to their means, for the religious and moral, intellectual, physical and social education of their children.

42.2: Parents shall be free to provide this education in their homes or in private schools or in schools recognised or established by the State.

The State shall not oblige parents in violation of their conscience and lawful preference to send their children to schools established by the State, or to any particular type of school designated by the State.

The State shall, however, as guardian of the common good, require in view of actual conditions that the children receive a certain minimum education, moral, intellectual and social.

The State shall provide for free primary education and shall endeavour to supplement and give reasonable aid to private and corporate educational initiative, and, when the public good requires it, provide other educational facilities or institutions with due regard, however, for the rights of parents, especially in the matter of religious and moral formation.

In exceptional cases, where the parents for physical or moral reasons fail in their duty towards their children, the State as guardian of the common good, by appropriate means shall endeavour to supply the place of the parents, but always with due regard for the natural and imprescriptible rights of the child.

The constitution also recognises the important role of the family in Article 41 as follows:

1.1°: The State recognises the Family as the natural primary and fundamental unit group of Society, and as a moral institution possessing inalienable and imprescriptible rights, antecedent and superior to all positive law.

1.2°: The State, therefore, guarantees to protect the Family in its constitution and authority, as the necessary basis of social order and as indispensable to the welfare of the Nation and the State.

A corresponding, but much shorter, provision was contained in the **Constitution of the Irish Free State**, as follows:
Article 10.
All citizens of the Irish Free State have the right to free elementary education.

All cases that have gone before the High Court have proceeded on the basis that Article 42 of the constitution has been subverted. As we will see upon reviewing the O'Donoghue case, a High Court justice made it clear that the State's constitutional duty to provide for the free education of children clearly extends to those children with special education needs.

The O'Donoghue Case

Paul O'Donoghue was an eight-year old boy from Cork when his mother, Marie, pursued a case through the High Court. Born 'normal', Paul contracted a viral infection that resulted in the contraction of Reye's syndrome that left him with a Severe and Profound General Learning Disability. Despite his mother's numerous attempts to seek educational provision for Paul, he was refused admission to most of the pre-schools and educational facilities that were available at the time. Mrs. O'Donoghue devoted most of her life to Paul's education and care, even when he was enrolled in clinics or hospital. When Mrs. O'Donoghue heard about the Peto Institute in Budapest, Hungary she took it upon herself, using her own funds, to seek assessment and treatment there for Paul. Feeling that Paul benefited from the treatment provided at Peto, Mrs. O'Donoghue eventually made arrangements for him to receive treatment in Cork, from a specialist from the institute.

At the same time she sought enrolment in a number of educational institutions without success or, in her view, without positive outcomes. Mrs. O'Donoghue and a group of concerned parents shared all the costs of Paul's education and treatment. The State provided no financial assistance to her. She concluded that Paul was being discriminated against and took action before the court.

The case was heard by Mr. Justice O'Hanlon and his rulings offer an interesting interpretation of constitutional and educational entitlements and rights at the time the case was heard (1992-1993). In addressing the constitutional issue Justice O'Hanlon wrote:

"I conclude, having regard to what has gone before, that there is a constitutional obligation imposed on the State by the provisions of Article 42.4 of the Constitution to provide for free basic elementary

Part Six

education of all children and that this involves giving each child such advice, instruction and teaching as will enable him or her to make the best possible use of his or her inherent and potential capacities, physical, mental and moral, however limited these capacities may be. Or, to borrow the language of the United Nations Convention and Resolution of the General Assembly – "such education as will be conducive to the child's achieving the fullest possible social integration [sic] and individual development; such education as will enable the child to develop his or her capabilities and skills to the maximum and will hasten the process of social integration and reintegration".

This process will work differently for each child, according to the child's own natural gifts, or lack thereof. In the case of the child who is deaf, dumb, blind, or otherwise physically or mentally handicaped, a completely different programme of education has to be adopted and a completely different rate of progress has to be taken for granted, than would be regarded as appropriate for a child suffering from no such handicap."

Justice O'Hanlon went on to note that the State has clearly lagged seriously behind in its provision for children whose needs are similar to Paul's. He added that in line with the conclusions of the Irish Commission of Inquiry on Mental Handicap, reporting in 1965, Paul might not be considered likely to benefit from educational services but that research and considerable progress in educational services had ensued since then and its findings were no longer relevant. (It was in response to this commission's report that Ireland established a system of special schools that were integrated into, and recognised as part of the National School system).

Justice O'Hanlon wrote about the education of children with Moderate General Learning Disabilities that was in place at the time, noting its emphasis on the child's

"…physical, intellectual, emotional, social, moral and aesthetic development to the maximum extent that it is possible to do so having regard to the degree of handicap from which the child is suffering."

He noted that children like Paul, with Severe and Profound handicap, differed only in terms of the severity of their needs and the degree of difficulty on the part of teachers in providing education to them. He also stated,

> "I believe that it has now come to be accepted that trained teachers and the school environment can make a major contribution to this process which cannot – with the best will in the world – be provided as effectively or as successfully by parents and family in the home".

Up to this point in his ruling Mr. Justice O'Hanlon referenced the constitution's emphasis on the family as the primary educator of the child, but noted that in exceptional cases such as that of Paul, the family cannot be expected to have the ability to provide for the educational needs of the children. He also noted that it is within the province of education to provide for certain educational needs of all children, referencing the State's curriculum at the time, for children with Moderate General Learning Disability. He emphasised the State's failure to provide any financial support to Mrs. O'Donoghue in her efforts to secure education for Paul. Mr. O'Hanlon's conclusions provide a basis from which we can assume several facts:

- People with Severe and Profound General Learning disabilities can benefit from education
- It is the duty of the State to provide this education
- The educational provision for these people addresses physical, intellectual, emotional, social, moral and aesthetic development
- These entitlements are rights enshrined in the Irish constitution

Justice O'Hanlon went further in his ruling. He outlined some of the specific aspects that characterise the educational provision of children like Paul and the extent to which these characteristics were not currently being met in Ireland. For example, he noted that the pupil-teacher ratio of 12 to 1 currently characteristic (at the time) for children with Severe and Profound General Learning Disability did not meet an international 'best practice' standard which was closer to 5 to 2 and that Denmark vastly exceeded even that small ratio. He also noted that the length of the school day which was in place in Ireland at the time for children like Paul was significantly shorter than an international best practice standard (which was a full school day). Finally, and importantly, he concluded that educational provision for children like Paul required a new approach, taking into consideration the following:

a) Age of commencement
b) Duration of primary education
c) Continuity of education

He made three specific points:

1. He noted that early intervention was important;
2. That education might well extend up to age 18;
3. That education should be continuous throughout the year and not characterised by the traditional long breaks associated with the education of children without these types of special needs.

In finalising his ruling, Justice O'Hanlon made a monetary award to Mrs. O'Donoghue.

The influence of the O'Donoghue case extends mostly to the non-legal aspects of special education provision because it did not result in any direct legislative imperatives. Developments subsequent to this ruling rendered it of little direct legal impact on special education provision. However Mr. O'Hanlon did give us an ethical, moral framework from which we can envision special education and assured that the Constitutional right to a free primary education was extended to those with even the most severe conditions. It is largely as a result of this case that classes for children with Severe and Profound General Learning Disabilities are in place throughout the country.

The Sinnott Case

Jamie Sinnott was born on 11 October 1977. At the age of four months, Jamie began to manifest symptoms associated with an Autistic Spectrum Disorder. His mother, Mrs. Kathy Sinnott, is the daughter of a physician who carried out an assessment of young Jamie and raised concerns about his developmental status. Several hospitalisations and assessments resulted in finding that Jamie had autism and Severe and Profound General Learning Disability, which is an associated condition among a significant percentage of children with autism. Mrs. Sinnott tried to access appropriate treatment for Jamie for a period of several years without success.

She then took Jamie to Chicago, Illinois for assessment at the paediatric unit of Loyola University Hospital. The assessment resulted in wide-ranging recommendations for occupational therapy, speech and language therapy and physiotherapy, along with educational interventions. These were carried out over a period of several months at the child centre of Michael Reese Hospital in Chicago. During this time, Jamie improved significantly in all aspects of his development. Mrs. Sinnott noted that in addition to improvements in Jamie's quality of living, there were significant effects on her own life quality in that she could sleep regular hours again, have a bit of a social life, and feel less stressed about her child.

At Christmas 1978, Mrs. Sinnott returned to Ireland along with copious notes, treatment recommendations, progress reports and clinical notations about Jamie's progress and treatment in Chicago. It was expected that this material would be used to provide appropriate educational and therapeutic interventions for Jamie in Ireland. Sadly this was not to be the case. Instead Mrs. Sinnott was subjected to advice based on discredited theories of autism and the lack of any meaningful coordinated service provision in Ireland. As a result Jamie began to regress. As a result of these difficulties Mrs. Sinnott returned to the Michael Reese Centre in 1981 with Jamie. While attending at this centre for a period of five months Jamie regained lost ground and continued to progress further in all areas of development.

From 1985 until 1988 Jamie received services at the Cork Polio clinic (COPE Foundation) and Mrs. Sinnott was not at all pleased with the results because no formal education was provided. In October 1988,

Part Six

Jamie participated in a course of education for the first time. A teacher who had trained in Hungary at the PETO Institute began working at the Cork Polio facility. This teacher provided one-to-one intervention with Jamie and five other children. Jamie made considerable progress in this programme for six months. However, the decision was made at Cork polio Centre not to continue this programme. Mrs. Sinnot and the other parents considered this decision to be a disaster for the children. Mrs Sinnot then continued the programme on her own.

Other places were tried but none had the success that had been experienced by the programme at the Cork Polio Centre. Things continued in a similar vein until 1996, when Jamie was able to receive services from a teacher who had knowledge of autism. He made great progress once again but he was reaching the age of eighteen and appeals had to be made to continue his education beyond this age. Further education was granted to Jamie but the specialist teacher who had helped him make good progress had left the facility by that time. In September 1997 Jamie was no longer eligible for educational services. At that time it was decided to place him in the Orchard, a facility of COPE. It catered to the needs of children with Severe and Profound General Learning Disabilities but did not specialise in the treatment or education of children with autism as well. (During the court case, several world-renowned experts in autism testified that the Orchard would at best be a beginning upon which a proper programme could be built.)

The sad case of the State's inability to provide appropriate educational treatment or therapeutic services for Jamie continues and there is little point in covering ground which is essentially similar to what has already been recited in the previous case. Mrs. Sinnott brought a case to the High Court and it was heard by Mr. Justice Barr who issued his ruling in 2000. He came to a number of conclusions that might be helpful to review, before coming to his actual ruling in this case:

1. A weakness in the administrative structure related to children with special needs was noted, particularly communication and coordination of services between those Ministries who hold responsibility for such

services (primarily Health and Education) and a lack of information among the administrators of the Department of Finance about the needs of those with special needs. Justice Barr wrote that these administrators "...appear to be insufficiently informed regarding the constitutional obligations of the State to the weak and deprived in society to enable them to assess realistically the degree of priority which should be attached to each such claim and the structure of priority which the State should devise in meeting its Constitutional as distinct from other non-Constitutional obligations."

2. The State has a long-standing reliance on non-governmental agencies such as charitable institutions and religious organisations in meeting the needs of its residents, especially the handicapped. Justice Barr concluded that though this practice is not in itself inappropriate, the State must take a hands-on approach, provide necessary funding, and take "...a positive role in the organisation, provision and supervision of services offered on its behalf..."

3. Poignantly Justice Barr wrote, "The sad history of Jamie Sinnott is an indictment of the State and cogently illustrates that it has failed to participate actively and meaningfully in the provision of appropriate services for him and those like him over the years."

4. Justice Barr cited the Department of Finance's failure to implement the rulings in the O'Donoghue case which established the student-teacher ratios of 1:6 for children with Severe and Profound General Learning Disabilities. He noted that internal documents indicated that this Department had failed in its duties to these children by not providing adequate funding to initiate the services these children require, such as appropriate speech and language therapy.

5. Justice Barr noted that there were hundreds of cases similar to that of Jamie Sinnott which are outstanding against the State and that this pointed to a fundamental problem requiring urgent attention.

After outlining his conclusions, Justice Barr paid considerable attention to the ruling in the O'Donoghue case by Mr. Justice O'Hanlon. This offered, in his view, the primary legal background for his own ruling. In a thorough restatement of his conclusion about each point raised in this case, Justice Barr's own words need to be recorded here. Although lengthy they remain a particularly pithy review, in fact an indictment, of the state of special education at the time, that can be used as a meaningful template to compare and contrast services, as they presently exist.

Mr. Justice Barr wrote:

"(i) In the course of his 23 years Jamie Sinnott has had no more than about two years of meaningful primary education and training provided by or on behalf of the State.

(ii) Jamie was a normal child in good health up to the age of four months who achieved the usual milestones until then.

(iii) At or about that age there was a profound change in him. He has been diagnosed as suffering from symptoms of severe autism and mental and physical dysfunction. It is not in dispute that he is and will remain severely or profoundly mentally handicapped.

(iv) A period of partial schooling (two hours or less per day) which had continued for about 18 months came to and end in July, 1997, the explanation given being that by reason of age, Jamie was no longer entitled to State education.

(v) From 23rd September, 1998 in consequence of pressure deriving from this litigation which was then heading towards trial, a place was found for the plaintiff at the Orchard, COPE which provides for severely or profoundly mentally handicapped people of about his own age. The group he joined also suffered from severe physical disablement and, unlike him, they were not ambulatory. None were autistic. The teacher in charge is unqualified and has no experience or training in autism and little training in dealing with the profoundly handicapped. There was not then and never had been a programme for Jamie's training or education. In the course of the trial an ITC programme for the plaintiff was hurriedly assembled for the first time. It was fundamentally flawed and was severely criticised by all of the experts, including Dr. Ware and Professor Mittler, the defendants' witnesses. Mr. O'Reilly, counsel for the defendants, has conceded that it is inadequate and must be replaced. Mr. Gerry Buttimer, the chief executive officer at COPE, conceded in evidence that his foundation has no-one available to it with experience of autism and no facilities for the education or training of autistic persons. The regime at the Orchard has been (in my view rightly) criticised as being wholly unsuitable for Jamie's education and training. The admittedly poor programme of instruction; the absence of essential therapies; insufficient facilities in that regard and the lack of any personnel who are trained in autism and its management amply bears that out. Alternative meaningful education and training for Jamie is an urgent imperative.

(vi) Jamie probably will need further education and training for the rest of his life. In my opinion those are not sustainable grounds for measuring his constitutional right in that regard in terms of actual age – particularly bearing in mind that the State has failed to honour its constitutional obligation to him for most of his life and in breach of duty has failed to provide such services for many years as a result of which he has suffered significant additional permanent harm. Jamie has thereby lost valuable time which has not only caused him unnecessary distress, but has also significantly damaged his prospects of fully recovering lost ground in his education and training. All of the experts agree that early intensive intervention with continuing education and training thereafter is of great importance if optimum results are to be achieved for those, such as Jamie, who are severely afflicted by autism and also physical and mental disablement.

(vii) The State has no alternative to the Orchard presently available to Jamie. There are tentative plans afoot at COPE and a centre for education and training of adults suffering from severe autism and mental disablement may eventuate there in two or three years time. Whether it will be adequately staffed with trained personnel and will be suitable for Jamie's on-going education, including job training which might lead to sheltered employment, remains to be seen.

(viii) Jamie has suffered substantial personal harm and damage by reason of the breach of constitutional duty of the State, its servants and agents, and its failure to honour its constitutional obligation to provide him with

Part Six

education, training and health care appropriate to his particular situation. This has been aggravated by persistent failure to honour the terms of the judgment of O'Hanlon J. in *O'Donoghue v. The Minister for Education and Others*, and also failure to provide the plaintiff, whose afflictions include severe autism, with vital ancillary services such as speech, occupational and physiotherapies.

(ix) The State's breach of duty includes:-

101. Failure to provide or have provided adequate primary education for Jamie Sinnott.

102. Failure to provide continuity of educational and other services for him.

103. Failure to provide necessary ancillary services, in particular speech therapy; occupational therapy; physiotherapy; and music therapy.

104. Failure to provide sufficient psychological and medical assessment and treatment for Jamie.

105. Failure to devise and operate an appropriate curriculum for Jamie's education and care.

106. Failure to devise, revise and keep in operation a viable programme for Jamie's education and training and to do so in consultation with his mother.

107. Failure to keep adequate records of his education, training and treatment. Failure to keep his mother adequately informed of her son's progress and of intended plans for his education and training.

108. Failure to collaborate with his mother in devising plans for his education and training.

109. Failure to recognise and respond adequately to his needs.

110. Failure to give him adequate training in personal care, hygiene and mobility.

111. Failure to address and provide instruction and treatment for his on-going drooling problem which is and has been a major difficulty for the plaintiff since infancy and a source of continuing distress.

112. Failure to provide him with any occupational training which might enable him to obtain meaningful sheltered employment.

113. Failure to provide for Jamie a teacher and other ancillary experts who are trained in autism and familiar with its problems.

114. Failure to establish and maintain reasonable co-ordination between the Orchard and Mrs. Sinnott.

115. Placing Jamie in an institution (the Orchard) which is unsuitable to his requirements and positively harmful to him by creating a climate for regression.

116. Failure to supervise adequately the services for Jamie Sinnott which the State contracted with the COPE Foundation and others to provide on its behalf from time to time.

117. Failure to take any adequate steps to ensure that such services were structured in a meaningful, appropriate way.

118. Failure to provide its contractors with the resources necessary to meet the constitutional obligation of the State to educate the plaintiff and to meet his special needs having regard to his particular disabilities as a person who suffers and has suffered from severe autism since the age of four months and major physical and mental handicap.

(x) Harm done to Jamie includes many bouts of anguish, physical and mental damage, depression and misery; also deprivation of the degree of happiness, well-being and human dignity which on the balance of probabilities he would have enjoyed if the State had provided him with the primary education and training which was his right.

(xi) Mrs. Sinnott has also suffered harm, loss and damage by reason of the failure of the State to honour its constitutional obligation to provide

Part Six

adequately for her son's education and training, all of which was a reasonably foreseeable consequence of the State's conduct in that regard."

Justice Barr awarded considerable financial damages to Mrs. Sinnott in his ruling and indicated that Jamie Sinnott was to be considered able to benefit from further education. These aspects of his ruling were to become the subject of the State's appeal to the Supreme Court in the case of Jamie Sinnott.

Sinnott Supreme Court Ruling

Space does not permit a thorough review of the Supreme Court ruling in the case of Jamie Sinnott. It is enough to note that it seems there is now a definitive ruling that the State's duty to provide for a free primary education does not exceed the age of eighteen. Additionally, the court rolled back some of the monetary awards to Mrs. Sinnott. However, the Department of Education indicated that it would honour Justice Barr's financial awards in full.

Other Court Cases

Too many to mention and review are the numerous cases brought before the High Court by parents of children with autism and other special education conditions. The most recent is the O'Carolan case, which is strikingly similar to both Sinnott and O'Donoghue in features. All the court cases take similar views of the State's failure to provide adequate and appropriate educational and related services to children with special education needs. Each case can be characterised by the difficulties in coordinating services that fall under the remit of several Ministries, the lack of appropriately trained teachers and related professionals to educate and treat these children, and the difficulties families face in attempting to secure appropriate service.

The O'Carolan case, resulting in a ruling in June 2005, featured testimony about the difference between "appropriate" and "adequate" services. Lewis O'Carolan has been deemed an acceptable client for a facility not unlike that which Jamie Sinnott was sent to so long ago, and that is viewed by the author (one of a number of experts to testify on behalf of

Part Six

Lewis in this case, all of whom agreed on this point) as being only the beginning of an appropriate service for children like Lewis.

The reader is urged to review the rulings in the O'Donoghue and Sinnott case as well as that in the O'Carolan case (not available at the time of writing this book) and use them as a template against which to compare existing services for children with special needs. Even a casual look at these cases and the written conclusions of each Judge involved will lead one to the opinion that there is a lot of ground to be covered before the State can honestly claim it is providing appropriate educational services to all children with special needs. The State continues to spend millions of euros defending itself against these court cases, money that could be put to better use improving programmes, increasing resources, and ensuring that appropriate provision is an everyday reality for all children with special needs, indeed all children enrolled in every level of Irish education.

Part Six

PART SEVEN

Reports of various Task Forces

In addition to numerous official policy papers, there have been two important unofficial documents published since 2000. These reports, prepared by committees established by the Department of Education and Science are not binding on the State but provide useful and important information that needs to be translated into further special education planning and development.

Autism Task Force

In October 2001, the report of the Task Force on Autism, entitled "Educational Support and Provision for Persons with Autistic Spectrum Disorders (ASD)", was released. This comprehensive review of autism in Ireland, from early identification to assessment to intervention made a number of recommendations that are relevant today. The task force recommendations covered policy, practice, early education and general educational issues. It is the general educational issues that are of interest for the purpose of this book. These recommendations include the following:

- "That the Department of Education and Science make available a range of approaches to meet the needs of children with an ASD including home-based services, mainstream and special schooling, and all related core therapies.
- A full range of resources to children with ASD's be provided in school settings.
- That funding be made available to investigate the efficacy of approaches, including those official departmental guidelines which relate to educating children with special needs and ASD's.
- That the Department issue a national directive underscoring the primacy of inclusion and enrolment in the least restrictive setting and that this directive include a requirement that school Boards must publish policies about inclusion and adaptations related to children with ASD's.
- That the Department set up a committee, to include parents, to agree guidelines and procedures to monitor progress.
- That schools adopt a whole-school ethos to support children with ASD's.

Part Seven

- That schools enrolling these children actively promote inclusion of these children.
- That the Department issue anti-bullying policies related to children with ASD's.
- That the Department assure a range of policies and supports for older children with Asperger's Syndrome or High Functioning Autism in schools and in communities.
- That the State put significant resources into early intervention and service programmes for young children with ASD's.
- That educational and clinical assessment be offered from the time of identification for each child with an ASD.
- That special classes for children aged five and under with an ASD be developed and located in mainstream *and* special schools.
- Re-organisation of the home tuition scheme to meet the needs of young children with ASD's.
- That children with ASD's be allowed dual enrolment in both special and mainstream schools.
- That statements of educational needs, IEP's, child and family support plans be available for young children with ASD's.
- Support should be provided as needed for all children with an ASD enrolled in a mainstream school.
- That a Visiting Teacher service for ASD be established and made available to young children in pre-school settings.
- That prescribed therapies and behaviour management supports be available to all children with ASD's.
- The provision for meaningful integration and inclusion for these children be written into school plans.
- That SNA's be provided with autism-specific training.
- That children with ASD's in primary school have access to the full curriculum.
- That schools review, at least annually, their provision for children with ASD's.
- That schools actively support and involve parents of children with ASD's in all aspects of their education."

Part Seven

The task force went on to make recommendations for second level schools that are specific to children with Autistic Spectrum Disorders. These recommendations should become familiar to all parents of children with a special education need enrolled, or soon to be enrolled, in a secondary school. They are far-reaching, insightful and visionary in the context of Irish education and could become the single most valuable source of information that can assist in the development of special education at secondary level. For this reason they are listed here in full. They are:

- That a flexible continuum of educational options be available.
- That there be a range of differentiation for children with ASD's, based on the two existing models, ASD and AD/HFA.
- Following completion of a multi-disciplinary assessment prior to leaving primary school, a Special Needs Organiser, working cooperatively with parents, identifies and secures an appropriate secondary placement for the child.
- That entitlements and supports in place at primary level be modified as needed and be available at secondary level.
- That any additional necessary supports be ring-fenced for the particular child.
- That support services be available to all students with an ASD, whether being educated at home, school or special setting.
- That the Department issues guidelines indicating the necessity of home-school communication and liaison for children with ASD's.
- That the Department urgently review the policy of attaching explanatory notes about special exam arrangements on the candidate's official certificate.
- That staffing of NEPS be increased to ensure availability of psychological services to students with an ASD.
- That the Department publish and circulate guidelines on bullying to secondary schools.
- That effectiveness of provision be monitored and published regularly by the Department.
- That second level placement options be based on the theory of Least Restrictive Environment, with local options having primacy.
- That all students with ASD's have access to an appropriate curriculum and accreditation route to match ability and direction.

- That a Visiting Teacher Service be available at secondary level to students with an ASD.
- That preparation and planning for post-secondary education be a part of the student's secondary education.
- That a review of the post-secondary transition takes place at age fourteen, or no later than the end of the Junior Cycle stage.
- That a multi-disciplinary assessment be undertaken of all students with an ASD as they approach completion of schooling.
- That the discrepancy between ability and achievement in students with AS/HFA be investigated with a view to identifying effective interventions to address this problem.
- That effective counselling services be provided to all students with ASD's at second level to reduce early school leaving.
- That all students with an ASD at second level have access to social, communication and life-skill training programmes.
- That the social and educational needs of students with ASD's at second level be prioritised by all who provide services to them.

These recommendations about secondary schooling are so important that I urge every reader with an interest in improving special education provision at second level to memorise them, recite them to school administrators and Department policy makers at every opportunity and to work tirelessly to see that they are adopted, adapted when necessary, and become policy in every secondary school in the State. They are the most useful, readily available source of recommendations appropriate for all students with special needs at second level and should not be perceived as relating solely to students with ASD's.

The Dyslexia Task Force

In July 2001, this task force issued its report. Presenting a wide-ranging review of special education issues related to children with Dyslexia, the report issued its recommendations in a manner consistent with the autism task force. We will outline those recommendations that most directly relate to the education of children with Dyslexia.

In creating its list of recommendations however, the task force recognised that while its remit was Dyslexia, the recommendations lend themselves to a great variety of special education conditions. They were also at pains to recognise that Dyslexia is but one of the Specific Learning Disabilities and framed their recommendations in the light of this knowledge. The recommendations of the Dyslexia task force include the following:

- The Department should ensure that information about these conditions is readily available to parents of children with Dyslexia and other Specific Learning Disabilities (SLD's).
- The Department should develop a plan for the development of additional special classes in mainstream schools for children with dyslexia.
- The number of hours of specialist teaching for children with dyslexia should exceed two and a half per week.
- A student-teacher ratio of 9:1 should be the norm in special classes for children with dyslexia, in mainstream schools and in special schools.
- SNA's should be assigned to special classes and special schools for children with dyslexia, on a needs basis.
- Enrolment in special schools or classes for children with dyslexia should extend beyond the current limit of two years, on a case-by-case basis.
- The development of a systematic approach to the transition back to mainstream schools should be developed.
- The Department should amalgamate all relevant circulars related to dyslexia and the specific learning disabilities and make them consistent.
- Identification of dyslexia should be based on a phased model that is outlined in the body of the task force report.
- The learning support teacher to be responsible for no more than thirty students in any given school term.
- Provision of a flexible level of learning support service at primary and secondary school level.
- Additional learning support hours should be provided (based on needs) on a temporary or permanent basis to those students whose achievement remains low as a result of dyslexia.
- No more than four or five schools should constitute a cluster for the provision of learning support services.
- Establishment of an appeals procedure.
- Ask the Director of Equality and the National Disability Authority to review the practice of appending explanatory notes on official State exam results.

Part Seven

- Inform candidates for State exams at least six months in advance if their applications for reasonable accommodations have been granted.
- The Department should commission the development of Irish language support and assessment materials.
- Consider the appropriateness of assessment and teaching materials for culturally different students.
- Provision of information and training about dyslexia at pre-service, and continuing education for all student and practicing teachers.
- The development of special schools for children with dyslexia as centres of excellence to provide resources and assistance to all teachers.
- Involvement of parents at primary and secondary school level in providing services to children with dyslexia.
- Schools to identify a teacher who will be responsible for coordination of services to children with dyslexia and SLD's.
- Involvement of parents in the assessment of children with dyslexia and SLD's.
- Schools should monitor the effectiveness of interventions for children with dyslexia.
- Class and subject teachers should assume major responsibility for children with dyslexia and other SLD's.
- Class and subject teachers to assume a major role in planning and assessment for children with dyslexia and SLD's.
- Parents, class teachers and specialist teachers should work together in the development of intervention plans.
- Learning support teachers should provide a pro-active role and advise class and subject teachers about appropriate support interventions.
- Class and subject teachers to participate in team meetings when support services are being considered for a student with dyslexia.
- Class and subject teachers should compile appropriate records of student's progress.

As has been stated above, these recommendations need to become part and parcel of every parent's knowledge and every teacher's knowledge. Where these sorts of reasonable recommendations are not in place, parents should demand that they be in place. Only concerted, tireless effort on the part of parents and concerned teachers will propel the special education system into higher and higher levels of effective provision.

PART EIGHT

The Author's Recommendations

So much has been produced, so many recommendations made, about special education in Ireland over the past decade that one feels humbled in offering additional suggestions. Be that as it may, I want to note my own ideas about special education, in no particular order of importance, but with the intention of making clear to my readers my own bias, which of course, lurks behind every word in the book. In that spirit I offer the following:

1. Special education in Ireland absolutely requires legislation that guarantees an appropriate education for every child with a special need; no exceptions, no escape clauses, and no equivocation.

2. No restrictions on the rights of parents, guardians, or students of majority age to access the courts to secure an appropriate special education should be part of any legislation.

3. Rights of meaningful appeal must be guaranteed in law and in policy to parents, guardians, and children of majority age.

4. Every child and adolescent with a special education need must be given an absolute guarantee that all the required resources to support their growth and development as human beings will be provided, regardless of cost, availability in the State, or inconvenience to Departments or systems.

5. No person who does not have a post-graduate diploma or degree in an area of special education and recent, relevant experience working with school-age children in an educational setting should lay claim to being a special educator. All teachers in the State, in order to qualify to work with children with special needs, should be required to undertake post-graduate training.

6. A Code of Ethics and Standard of Practice related to special education must be developed for teachers and must become part of the employment contract of every primary and secondary school teacher.

7. The Department of Education and Science must recognise the requirement for a post-graduate education to qualify as a special education teacher and must provide a financial incentive to secure this education in the form of grants to continue education and salary increments that recognise the awarding of the appropriate post-graduate diploma or degree.

Part Eight

8. All teachers currently employed in a special education post who do not hold a post-graduate diploma or degree in special education should be given five years to secure one and the necessary supports to facilitate this requirement should be put in place by the DES. Failure to earn such a credential should result in removal from special education and return to the mainstream classroom.

9. The provision of supportive services for children with special education needs should fall under the remit of the National Council for Special Education in order to avoid confusions, poor co-ordination and lack of service. (This means that OT, PT, SLT related to educational issues must come under the direct control of the DES).

10. All teachers must recognise and manifest the requirement to engage in life-long continuing education, particularly as it relates to children with special needs.

11. The Colleges of Education must include intensive, mandatory course work in special education at pre-service level. The colleges must continue to develop more specialist post-graduate and continuing education options for practicing teachers.

12. The Colleges of Education must begin to offer programmes for parents that provide them with the essential information about special education provision, conditions and family related issues.

13. The staff of the Colleges of Education must engage in continuing research about the efficacy of special education interventions and programmes, and publish the results.

14. The staff of the Colleges of Education must become active critics of the education system as a whole and should publish and disseminate their critiques nationally and internationally.

15. All staff working under the umbrella of the National Council for Special Education must be able to demonstrate a post-graduate diploma or degree in special education and recent, relevant experience in a special education setting within the recognised primary or secondary sector (work experience in medical or institutional settings outside of the approved primary or secondary sector should not be a recognised special educational credential).

16. All teachers must be required to engage in continuing professional development, to be reviewed and certified by the Teaching Council every five years.

17. Financial recognition of advanced post-graduate diploma or degree status must be provided by the DES to every qualified teacher.
18. Class size must be reduced to a reasonable level (Infant classes-17, Junior classes-21, Senior classes-25) to permit and facilitate inclusive special education practice.
19. School principals must be required to demonstrate, in practice and action, a clear commitment to inclusive special education, such demonstration to be reviewed every five years by the DES.
20. Special education provision for children with low incidence, high intensity, conditions must proceed on the basis that it is a year-long right with appropriate summer and holiday programmes on offer free of charge to these children and their families.
21. The psychological services should be extended to assure that every primary and secondary school has access to an educational psychologist at least one day every week.
22. A meaningful and enforceable system of child protection must be put in place to assure that all families meet the educational needs of all their children.
23. Teachers must be given adequate time for planning, meetings, and review of special educational provision during the school week. This will require a review of the teacher's duties and perhaps the addition of appropriately educated subject specialists at primary level. It is impossible to implement any special educational plans currently in place without adequate time for team discussions.
24. Parents must be involved at every level of special education provision, without exception, and their involvement should be meaningful.
25. Teachers must be trained to work in genuine, meaningful partnership with parents, especially parents of children with special needs. This training should begin at pre-service level and extend to continuing professional development.
26. Parents must recognise once and for all that teachers are not the enemy, recognise the goodwill that exists among so many teachers, and begin to work cooperatively with them.
27. Parents must recognise that it is the teacher who knows best how to teach.
28. Teachers must recognise that it is the parent who knows the child best.

Part Eight

29. Teachers in the special schools must begin to reach out to their mainstream colleagues, offer information and advice, and share teaching methodologies on a regular basis.
30. Teachers in mainstream school should begin to seek advice and input from their colleagues in the special schools.
31. Special schools for children with Mild General Learning Disabilities should be phased out over the next five years, their staff to become specialists working with other special and mainstream schools.
32. Special schools for children with Moderate General Learning Disabilities should be phased out over the next ten years, their staff to become specialists working with other special and mainstream schools and classes.
33. Educational psychologists should recognise that some special schools offer the best possible provision for some children with some special needs conditions, and provide parents with meaningful choices about where a child will be educated.
34. Every member of the Inspectorate should be required to engage in continuing professional development on an annual basis in the area of mainstream special education.
35. Students enrolled in Higher Diploma programmes to qualify as secondary school teachers should be required to take intensive course work in special education and demonstrate the ability to differentiate their teaching for children with special needs before gaining official recognition as a teacher.
36. Special education services at secondary level must be improved as a matter of urgency and of the highest priority.
37. A moratorium on the issuing of Green Papers, White Papers, Reports of Task Forces and Commissions, and any other unenforceable publications should be declared until the recommendations previously written have been adopted or adapted and put into practice.
38. The rights of gifted and exceptionally able children should be given equal status in provision to those of children with other special education conditions.
39. The National Parents Council-Primary should change its membership eligibility to permit individual parents to join.
40. Schools, primary and secondary, should become covered under the Freedom of Information Act immediately.

APPENDICES

APPENDIX ONE

Understanding Assessments and Academic Testing

Educational psychological assessment is a formal procedure undertaken individually between a psychologist and a child (or any person). After building up rapport and making the child comfortable, the psychologist will administer a test to the child. There are many tests that can be administered so I will just introduce and explain the most commonly used tests. First though, it is necessary to discuss some general principles of testing.

Tests do not tell anyone anything! They do however yield data and information that then must be intelligently consumed by the qualified assessor. The information gathered from a test must fit into the picture of the whole child's life, background, family dynamics, learning and schooling history, motivation, health history and a thousand other variables. Anyone who takes the simplistic view that a test provides an answer that can be used to definitively unlock the riddle of a child's learning problems is seriously mistaken.

What is Intelligence?

Arguments about the nature of human intelligence and what comprise it are centuries old. We have looked at one particular psychological test that is based on a model of intelligence that conceives it as a combination of verbal and non-verbal skills. This model feeds into the common-held understanding that being intelligent means you will 'be good at reading and maths, you will perform well in comprehensive examinations and you will necessarily perform well in school and get into the university programme of your choice'. This is obviously a narrow model and a dangerous assumption – it's one that is being challenged vigorously from many fronts today.

IQ is probably the most commonly understood, and at the same time misunderstood, concept about human intelligence. The problem with IQ scores is that they are far too easy to misinterpret and lead quickly to assigning people into the general categories of 'smart' or 'limited'. An IQ is nothing more than a mathematically derived formula to quantify various test scores. There has been a lot of research into IQ and it has been discovered that it is a good predictor of one thing, and one thing only – success in school! This is particularly true of the Wechsler model of IQ.

As stated earlier, there are other models, and a popular one is that developed by Howard Gardner*, which is known as Multiple Intelligence theory. Gardner was dissatisfied with the major model of intelligence and the type of educational structures that have resulted from this model. Gardner has stated, "I want my children to understand the world, but not just because the world is fascinating and the human mind is curious. I want them to understand it so that they will be positioned to make it a better place. Knowledge is not the same as morality, but we need to understand [morality] if we are to avoid past mistakes and move in productive directions. An important part of that understanding is knowing who we are and what we can do...".**

Gardner believes that every human being possesses a number of different capacities, all as a result of brain system functioning, which can be called 'intelligences'. For Gardner, there are at least seven different types of intelligence and our inability and difficulty in recognising them is a result of the way we educate people, relying mostly on words and numbers. Gardner lists his seven intelligences as:
- **Linguistic Intelligence** – facility with words, reading, speaking.
- **Logical-Mathematical Intelligence** – facility with numbers, logical-sequential thinking.

* Gardner, H. (1983). *Frames of Mind*. London: Fontana Press.
** Gardner, H. (1999). *The Disciplined Mind: Beyond Facts And Standardized Tests, The K-12 Education That Every Child Deserves*, New York: Simon and Schuster.

- **Interpersonal Intelligence** – being sensitive to and understanding other people.
- **Intrapersonal Intelligence** – knowing oneself, one's beliefs, attitudes and values.
- **Musical Intelligence** – the capacity to respond to music or perform music.
- **Bodily-Kinaesthetic Intelligence** – ability to move, athletics, dance.
- **Artistic Intelligence** – responding to or creating visual or plastic art.

Since creating his first list of seven intelligences Gardner has continued his research and now believes there are sufficient grounds for adding one other:

- **Naturalist Intelligence**, which he defines as enabling "…human beings to recognize, categorise and draw upon certain features of the environment."

Gardner continues to investigate the nature of human intelligence and is considering the possibility that he can add to his list the following:

- **Moral Intelligence** – a concern with those rules, behaviours and attitudes that govern the sanctity of life – in particular, the sanctity of human life and, in many cases, the sanctity of any other living creatures and the world they inhabit.
- **Existential Intelligence** – a concern with 'ultimate issues'.
- **Spiritual Intelligence** – exploring the nature of existence in its multifarious guises.

According to Gardner, every human being has capacities wired into their brains that are manifested in these intelligences to one degree or another. In other words, we are all intelligent – it's just that we show our intelligence in different ways. I have oversimplified this theory to a great degree but the point I wish to draw to your attention is how much we can underestimate certain children when we conceive of intelligence merely as an IQ figure obtained from a particular test.

This 'obsession' with IQ testing can lead to low expectations on the part of educators when the test results are low; low expectations quickly translate into poorer teaching methods, less reinforcement in the classroom for the child and therefore lower performance on the part of the child. The stakes are high in the IQ race and the winners are more often than not created by those who teach them than by any so-called natural intellectual endowment. Think of it this way; if a person has an IQ of 185 but possesses no ability to understand himself or others, what sort of a life will he lead? Generally speaking the answer will be a life of frustration, wrong choices, unhappiness in love and relationships, and constant disappointment.

Gardner's theory has its critics and is not universally accepted as an alternative model of human intelligence. But whatever concerns arise about it and from it there is no disputing the fact that Gardner is responsible for brining to the fore the issue of intelligence and expanding our understanding of what it is.

Intelligent tests require intelligent testers

Put simply, intelligence testing requires intelligent testers. Additionally, tests in themselves are not useful instruments to classify children into special education categories. I have seen far too many children with autistic spectrum disorders given tests of intelligence with the results indicating that their intelligence is significantly impaired – yet anyone working with the child, or any family member, can relate poignant stories of the child's keen mind and different ways of thinking that clearly demonstrates their intelligence. I am therefore wary of intelligence tests when they are relied upon to provide *the sole answer*, solution, and source of information used to provide special education services.

The information gathered from a test falls into some general categories. If the purpose of the test is to assess intellectual skills (often referred to as *cognitive skills*), the information gained should shed light on most of the following:
- Verbal skills
- Non-verbal skills
- Attention and concentration

- Visual memory
- Auditory memory
- Short-term, long-term and immediate recall of visual and auditory information
- Social judgement
- Social comprehension
- Hand-eye coordination skills
- Perceptual organisation skills (orientation in space and time)
- Abstract reasoning, both verbal and visual

If the purpose of the assessment is to investigate alleged behaviour or emotional problems, then in addition to the above, the information gathered should shed light on:
- Frustration tolerance
- Impulse control
- Anger management
- Coping skills
- Interpersonal judgement
- Stress tolerance
- Anxiety issues
- Fears and phobias
- Unusual thoughts or ideas/beliefs
- Knowledge of right from wrong
- Social problem-solving skills
- Motivation for schooling
- Preoccupations and obsessions
- Mood (emotions of short duration)

In general, the more information one is able to gather from assessment, the greater the possibility of putting together an intelligent formulation that helps everyone involved in the life of the child to understand the child more comprehensively. The written assessment should clearly provide insight into the referral question and all related information necessary to understand the child. It should be easily read by a teacher or parent, should not contain jargon or scores that can not be understood by all who read it and should outline specific recommendations for educational and support services and strategies necessary to enable the child to benefit from their educational programme.

Wechsler Intelligence Scale for Children

The most common assessment instrument used by psychologists is the Wechsler Intelligence Scale for Children and will therefore be the one we look at. In using this test as an example, you will hopefully get a reasonable picture of how an assessment is carried out. The Wechsler test is, essentially, a test of intelligence. It has been in use for over fifty years and has been revised numerous times to keep it up to date.

The test is divided into two sections with each section containing a number of subtests. The two broad sections of the test are the
• Verbal Scale
• Performance Scale

Successful completion of any item on any of the Verbal subtests requires a *verbal response*. On the Performance subtests, the person must *do something* in response to a question or task. When the entire test has been administered, the assessor calculates what is called a Composite Score, a score that takes into account both sections. Because it is a test of intelligence, the test scores obtained are called IQ scores and you will see the results stated in this format:
• Verbal Scale IQ
• Performance Scale IQ
• Full Scale IQ (the composite score)

The Full Scale score, according to the standard interpretation, indicates the level of a person's intelligence. A Full Scale score in the range of 90 to 110 is considered average; the person can be said to have average intelligence.

In addition to looking at the Full Scale score, the three scores, (verbal, performance and full scale) can be compared against one another. What is expected in most people is that the three scores will cluster close enough together to indicate that the individual's verbal and performance skills are evenly developed. When there is a large difference between the two subtest scores (verbal and performance), it may indicate learning problems.

This is as good a time as any to introduce the reader to some of the common terms used in assessment:
• Percentile scores
• Reading age scores
• Standard scores

Children are frequently referred for assessment after reading or maths tests have been administered to the entire class. The most common whole-class tests in use are *the Drumcondra tests (reading) and the Micra T test (mathematics).* A child's results in these tests are reported in what are called **percentile scores**. A percentile score indicates where a child stands in comparison to a sample of children in his or her own age, on a given task. A score at the 50th percentile on the Drumcondra test means that the child is well within the middle range (49 children out of 100 score higher, 49 score lower).

Some tests yield what are called **reading age scores**. Reading age scores do not yield significant information, their use has been criticised and has been discouraged in the learning support teachers' written guidelines. A two-year difference in reading age in 5th class may not be terribly significant.

Standard scores are also frequently reported following assessment. The average standard score is 100, which is at the 50th percentile, meaning the child's score isn't significantly different in that test than other, same-age children. Standard scores must differ from one another by about fifteen points in order for the difference to be of any real significance. About two-thirds of all children have standard scores on a test that are between 85 and 115, that is, the 16th percentile and the 84th percentile (see table below). Scores in this range are not particularly noteworthy (there are exceptions to this, which will be presented when we explain tests of children's intelligence).

The following chart will be helpful in translating standard scores, scale scores, standard deviations, and percentile scores into understandable and meaningful information. Standard deviations tell us how much confidence we can place in a given score. Any time a test is administered there will be a certain *range* of scores obtained that don't have any significance in the actual test results. In psychological assessment the usual standard deviation of significance is three points or more. So if a child measures 12 points on a test and 11 on another there is no real significance to this difference. Although an oversimplification it is helpful to consider the standard deviation in scores to determine whether or not a strength or weakness is actually present upon assessment.

Std. Score	Scale Score	Std. Deviation	Percentile
145	19	**+3**	>99
140	18		>99
135	17		99
130	16	**+2**	**98**
129			97
128			97
127			96
126			96
125	15		95
124			95
123			94
122			93
121			92
120	14		91
119			90
118			88
117			87
116			86
115	13	**+1**	**84**
114			82
113			81
112			79

111			77
110	12		75
109			73
108			70
107			68
106			66
105	11		63
104			61
103			58
102			55
101			53
100	10	**0**	50
99			47
98			45
97			42
96		39	
95	9		37
94			34
93			32
92			30
91			27
90	8		25
89			23
88			21
87			19
86			18
85	7	**-1**	**16**
84			14
83			13
82			12
81			10
80	6		9
79			8
78			7
77			6

76			5
75	5		5
74			4
73			4
72			3
71			3
70	4	**-2**	**2**
65	3		1
60	2		<1
55	1	**-3**	<1

The most useful scores to interpret for common sense purposes are therefore percentile scores. I recommend you ask for percentile scores when test results are being reported. Most importantly, do not expect reading- or mathematics-age scores to be useful for educational planning or for reviewing the effectiveness of educational interventions. We will refer to percentiles again throughout this section.

Standard Scores

The Verbal Scale, Performance Scale, and Full Scale scores are all Standard Scores. Previously I stated that standard scores all have 100 as their average, with the range of average being from 90 to 110. About two-thirds of all children will score between 85 and 115 on these three scales and scores within this range are not highly significant.

At the risk of getting bogged down in too much information, it's worth having a more detailed look. For example, let's take a look at the Verbal Scale. The subtests that are administered are in bold and I have included what they are trying to assess:

- **Information**: factual knowledge, long-term memory, recall.
- **Similarities**: abstract reasoning, verbal categories and concepts.
- **Arithmetic**: attention and concentration, numerical reasoning.
- **Vocabulary**: language development, word knowledge, verbal fluency.

- **Comprehension**: social and practical judgment, common sense.
- **Digit Span**: short-term auditory memory, concentration.

On the Performance Scale, the following subtests are administered (bold) and what they are trying to assess is indicated:
- **Picture Completion**: alertness to detail, visual discrimination.
- **Coding**: visual-motor coordination, speed, and concentration.
- **Picture Arrangement**: planning, logical thinking, social knowledge.
- **Block Design**: spatial analysis, abstract visual problem solving.
- **Object Assembly**: visual analysis and construction of objects.
- **Symbol Search**: visual-motor quickness, concentration, persistence.
- **Mazes**: fine motor coordination, planning, following directions.

An example will help illustrate the fine points of interpreting this test. Suppose Patricia is referred for an educational psychological assessment, having progressed through Stages One and Two (see page xii). The Wechsler test is administered and she obtains the following results (this is a crude example for illustrative purposes and the numbers are not meant to be accurate representations of what a real test profile would look like). Individual subtest scores range from a low of one to a high of nineteen. Remember that differences of three points or less between them are not particularly significant. When the difference exceeds three points it may indicate a difficulty with the underlying brain processing tasks that were described above.

Verbal Scale		Performance Scale	
Information	8	Picture Completion	9
Similarities	3	Coding	10
Arithmetic	9	Picture Arrangement	11
Vocabulary	9	Block Design	2
Comprehension	18	Object Assembly	9
Digit Span	9	Symbol Search	8
		Mazes	14

Using the conversion tables available in the Wechsler test manual, the results of these subtests yield the following scale scores:

Verbal Scale IQ	109
Performance Scale IQ	113
Full Scale IQ	110

Patricia is in the average range, right? Looking at the three Scale scores, you would think so. But if we take a closer look at the individual subtest scores, something interesting comes into view. On two subtests that assess abstract thinking (Similarities and Block Design), Patricia's subtest scores are quite low. Subtest scores have an average of ten and there is little significance in a variation of three. However, Patricia's score of 2 on Block Design and 3 on Similarities indicates a real weakness in abstract thinking, verbally and non-verbally, despite her average intelligence. This weakness may well indicate learning problems.

I described percentile scores earlier. These scores help us to compare a child's test results with those of other, same-age children. Let's see how Patricia compares with other girls her age by looking at the percentile scores that correspond to each of her scores above, as follows:

Verbal Scale		**percentile**	**Performance Scale**		**percentile**
Information	8	25	Picture Completion	9	37
Similarities	3	1	Coding	10	50
Arithmetic	9	37	Picture Arrangement	11	63
Vocabulary	9	37	Block Design	2	1
Comprehension	18	99	Object Assembly	9	37
Digit Span	9	37	Symbol Search	8	25
			Mazes	14	91

The results of all these subtests yield the following scale scores:

		percentile
Verbal Scale IQ	109	73
Performance Scale IQ	113	81
Full Scale IQ	110	75

Taking a look at the percentile scores tells us more about how Patricia compares to children her own age.

Now, let's suppose that Patricia was initially referred because she was having considerable difficulty learning to read. I was at pains to point out in the earlier section that the assessor must take into account all the factors that might result in Patricia's difficulty, before drawing conclusions. She may have had health problems which caused her to miss one-third of the school year over each of the past several years; what if her parents were members of the Travelling Community and moved her from school to school five times each year? What if, for the past two years she has had three different teachers, as a result of staff illness, and two of them had no teaching qualification? There may be personal issues (family bereavement etc) that may have relevance. Any of these factors, and more, could be the real cause of Patricia's reading problems. The assessor will have to take everything into account and put it together in a way that makes sense to all.

What I am saying here is that there are a great many factors which can account for the scores obtained and that it is the responsibility of the examiner to be sure the results are an accurate picture of the child's intellectual skills and not an artefact of other influences which mask the true skill levels.

It is only possible to make full sense of test scores if they are stated in full in the assessment written report. It is often the case that the psychologist will only report a range of scores, for example, "Verbal IQ: Average Range", "Performance IQ: Borderline Range" This sort of report writing can raise more questions than answers because sometimes the numbers are at the fringes of a range. For example a score of 90 and a score of 109 are both within the 'Average' range but are both at the extreme range, with one Low Average and one High Average. Without stating the exact numbers, it is impossible to get an accurate picture of the child's level of abilities. I suggest that parents request the complete test data, (the actual numbers themselves) – it will be a useful means to compare results if an assessment is re-administered sometime in the future.

Scale Scores

This brings us to the range of *scale scores* and what they represent. The Wechsler test is supposed to be a test of intelligence and for these purposes, the three scale scores that are calculated correspond to a range of intelligence 'category', from Gifted to Learning Disabled:

Scale Score IQ	Intelligence Range/Special Ed Category
130 and above	Exceptionally Able/Gifted
90-110	Average *(not a special education category)*
70-79	Borderline General Learning Disability
50-68	Mild General Learning Disability
35-49	Moderate General Learning Disability
Below 35	Severe/Profound General Learning Disability

You might wonder what happens to those children whose scale scores fall between 80-89. The short answer is that they are not generally eligible for special education services; if their reading of mathematic ability is below the 10th percentile they will be looked after by the learning support teacher. If not, they are deemed to be doing as well as other children and will not receive any specialist assistance.

As has been stated earlier, *observations* are a helpful source of information but it must be remembered that all observations are subjective (liable to be distorted by hidden bias and differing levels of tolerance for learning differences and differences in behavioural skills). In any assessment the sole reliance upon observation and teacher-made tests is inappropriate. Assessment instruments that generate quantifiable data are a necessary part of the assessment process.

Academic testing

IQ testing is only one part of the assessment equation. In addition to taking a look at intelligence and how a child's brain is processing information, it is critically important to take a specific look at academic achievement. Academic achievement in reading and mathematics can be assessed in a variety of ways. Teachers create tests in classrooms and make observations of children's learning. These are all *subjective* and though they yield useful information, there also needs to be an *objective* assessment of academic skills, particularly reading and mathematics. Comparing and contrasting the subjective with the objective gives us valuable information about a child's learning skills and achievement.

There are a great many tests of reading and mathematics; rather than list them, I will describe what academic testing consists of and how it must be compared with intelligence testing. The assessment of academic skills is a special education assessment. It differs from a psychological assessment in that it will generate no information about a child's level of intelligence. It is confined only to measuring, with some accuracy, the academic skills a child has achieved to date – it is measured on an individual basis, with child and teacher working together in a private place by themselves. This sort of assessment is absolutely necessary in order to discover if the child has a particular type of Specific Learning Disability. It is also required in order to be able to measure the success of a special education programme for the child.

Academic achievement is subject to a great many variables as has been already stated. So, a test of academic achievement, like a test of intelligence, must be thoughtfully and carefully considered. Remember, tests don't tell you anything – they give you evidence and data that must be interpreted in light of what one knows about the whole child. There are academic tests for just about any content area one can consider, but for the purpose of special education assessment, the two most common areas that are assessed are *reading and mathematics*. Let's look at reading as an example of special education assessment.

How do we learn to read and how do we assess reading?

Reading is the final culmination of a large number of brain-related skills. Human beings are not biologically programmed to read. We are programmed to speak, to utter sounds, but not to read. Reading is something we impose on the human brain and it happens in a rather predictable sequence. First, the brain must be able to recognise the sounds of speech that make up words; these are called 'phonemes' ('phonics' is now a widely used methodology used when teaching children to read). Then the brain must begin to bunch these sounds together into words, while at the same time having a visual image of the word that is being heard or spoken. 'C' is a phoneme, so is 'a' and so is 't'. C-a-t soon enough become blended together to make a word that symbolises the animal we are familiar with and have an image of in our brain.

Sooner or later the brain has to learn that the sound of 'c' is also associated with a funny-looking shape called the letter 'c', known as a 'grapheme' and on and on it goes until one day we are able to hear and understand, say and understand, read and understand the word 'cat'. Many different brain systems are working together in learning to read.

All of the above makes the assessment of reading a tricky piece of business! The assessor must discover not only what level of reading skills the child possesses (in percentile terms, not reading-age terms) but must also discover just where any breakdown or difficulty is occurring:

- Is it at the level of being able to discern phonemes (the letter sounds)?
- Is it at the level of deciphering the graphemes (the shape that represents each letter) and relating them to phonemes?
- Is it at the level of being able to retain in memory what has just been read?
- Is it at the level of understanding what has been retained in memory, known as 'reading comprehension'?

Anything can go wrong at any stage of the process and the assessment of reading (and the same for mathematics), must take the entire process involved in mastering those particular academic skills into account.

When tests of reading or mathematics have been administered, they yield test scores, the same way that tests of intelligence yield scores. Let's stick with the example of Patricia, whose Wechsler scores were discussed earlier. Let's imagine that Patricia was given an academic test of reading, which broke down into several areas, each with its own score:

- Ability to discern phonemes
- Ability to read single words
- Ability to remember what was read and recall it
- Ability to comprehend what was read and repeat it by saying what it was about

Again, the most useful scores that can be reported are standard scores and percentile scores, because they can be compared and contrasted easily with other tests, say psychological tests. Patricia's scores on the test look like this:

	Standard Score	**percentile**
Phoneme Awareness	108	70
Word Reading	102	55
Remembering	98	45
Comprehension	62	1

We can compare these scores with other scores and quickly see that with her average intelligence, Patricia is able to discern phonemes, read single words and recall sentences within the average range. However, her ability to comprehend what she has read is well below average. When we look back at the intelligence test results, we notice that both verbal and non-verbal abstraction skills were quite poor. This helps us understand the reason for Patricia's poor comprehension – she is unable to draw conclusions and make generalisations about the material she has read. It is her inability to think abstractly that makes these two things difficult/ impossible for her.

What is important to look for is any real difference between full-scale intelligence test scores (Wechsler) and achievement test scores (academic testing of reading or mathematics). We must always bear in mind that full-scale test scores do not always reflect the true intelligence. Analysis of the individual standard scores and sub-section scores tells an interesting story to the assessor. Relying solely on the percentile score just tells us where Patricia stands on a task in comparison to her same-age peers. But comparing and contrasting the individual scores with one another tells a much more comprehensive story about the possible causes of her reading difficulties.

In conclusion, it is important to note that in order to make a clear decision about whether or not a special education condition is present, one must take into account both intelligence test scores and academic test scores. Failing to do so can lead to inappropriate placement into special education services or failing to provide special education when it is necessary to do so.

Assessment in special education must therefore always consist of at least two parts:
1. assessment of intelligence and cognitive skills (performed by the psychologist).
2. assessment of academic skills (performed by the psychologist and the special education teacher).

There may well be need for other types of assessment by other professionals in individual cases:
• A child or adolescent psychiatrist
• An occupational therapist
• Physiotherapist
• A speech and language therapist

Assessments must be individualised for the child and may take the form of any of the following methods:
Criterion-referenced Assessments
These are tests that include items that are *directly related* to what the child is supposed to learn. The goal is to obtain a description of the specific knowledge and skills the child can demonstrate.
Norm-Referenced Tests
A test or assessment designed to provide a measure of the child's ability on a particular task, as compared to other children the same age, class and gender.

Curriculum-based Assessments

This is a way of monitoring a child's performance through the existing subject content. The material used to assess the child's skills is often taken directly from the course content itself.

Checklists and Rating Scales

These are a form of criterion-referenced assessment. A teacher completes a series of questions, sometimes answered 'yes' or 'no' and the results are compared against those of children whom the test designer knows do not have learning difficulties.

Portfolio Assessments

Portfolios are a systematic collection of a child's work over a period of time used for the purpose of tracking progress in a particular content.

Language Sample Analysis

Often used by speech and language therapists, this form of assessment is a simple record of what a child says. The sample is then analysed for errors in grammar or sentence construction, such as how the child changes a singular to a plural.

Functional Assessments

This is a way to assess a child's behaviour. Looking at problem behaviour from this point of view, the assessor is trying to determine what the purpose of the behaviour is for the child, what function it serves. This information can then be used to modify the environment to correct the behaviour, or to use other strategies to help the child's behaviour improve.

Play-based Assessments

Play-based assessments are often used with young children or children with autistic spectrum disorders or those with general learning disabilities. The child is engaged in play and the assessor makes careful notes while observing the child's play patterns.

Ecological Assessments

An ecological assessment looks at the environment in which the child is being educated and makes decisions about how appropriate it is for the particular child in question. Ecological assessment also focuses on the people in the learning environment, not just the environment itself.

These approaches yield rich information about children, are especially important when assessing students who are from culturally or linguistically diverse backgrounds and therefore, are critical methods in the overall approach to assessment.

Observation

Observation is a useful method of assessment and can take many forms as follows:

- **Teacher observation** – teacher can observe a child and make careful notes on any area of development.
- **Time-Sampling Recording** – monitoring and tracking *everything* that happens over a given period of time.
- **Duration Recording** – documents the amount of time a child is engaged in a particular type of behaviour.
- **Event Recording** – recording a particular 'event', such as a temper tantrum. Of course, the event can be a positive one as easily as it can be a negative one!
- **Anecdotal Records** – a written record of a specific event that catches the teacher's attention in the classroom or on the yard.
- **Interview Based Assessment** – talking with, and listening to, the child. A written record is often kept and analysed or compared with another interview at a later time.
- **Functional Behaviour Analysis** – looks at the purpose of a child's behaviour in a particular context. For example, the purpose of attention-seeking behaviour may be quite different when it occurs on the yard from when it occurs in the classroom.

Let me stress again that tests alone will not give a comprehensive picture of how a child performs or what he or she knows or does not know. Evaluators need to use a variety of tools and approaches to assess a child, including observing the child in different settings, to see how he or she functions in those environments. They must interview individuals who know the child to gain their insights, and must test the child to evaluate his or her competence in whatever skill areas appear affected by the suspected disability. In addition, they must pay attention to any areas of particular strength.

Parents of children with medical or mental health problems may also have assessment information from sources outside of the school. Such information would need to be considered in tandem with that from the school's own evaluation team in making appropriate diagnoses, placement decisions and instructional plans.

A word of caution about observation

While observations will yield much useful information about the child and his or her environments, there are a number of errors that can occur during observations that distort or invalidate the information collected:

- The observer must record accurately, systematically, and without bias. If his or her general impression of the child influences how he or she rates that child, in regards to specific characteristics, the data will be misleading and inaccurate. This can be especially true if the child comes from a background that is different from the majority culture. In such cases, it is important that the observer has an understanding of, and a lack of bias regarding the student's cultural or language group. Often, multiple observers are used to increase the reliability of the observational information collected.
- All observers should be fully trained in how to collect information using the specific method chosen (e.g., time-sampling using a checklist) and how to remain unobtrusive while observing and recording, so as not to influence the child's behaviour.
- It is important to observe more than once, in a number of situations or locations, and at various times.
- Data must be integrated with information gathered through other assessment procedures. Decisions should not be made based upon a narrow range of observational samples.

Additionally, it will be important for the multi-disciplinary team responsible for educational planning to take into account the following:

- Who will make the observation;
- Who or what will be observed;
- Where the observation will take place (observing a range of situations where the student operates is recommended);

- When the observation will take place (a number of observations at different times is also important); and
- How the observations will be recorded.

It is good practice for parents to ask specific questions about the type of assessment or observation that will be/was used to pinpoint a child's level of performance on a task or behaviour. It is reasonable to ask the assessor how much practice they have in using the particular assessment method, how many assessments like this they have done, why they chose that particular method and how valid or reliable they think it is.

APPENDIX TWO

Support Groups, Websites and Other Sources of Information

Support Groups

There are a large number of parent support groups in Ireland that are a tremendous source of information for parents and teachers. Do not be afraid to ring them up and ask questions. Remember that the only foolish questions to be asked are the ones you are afraid to ask. If they can't answer the question ask them to refer you to someone that can. Among the helpful Irish support groups are the following:

Caint
A Support Group for Speech and Language Impaired Children
10 Bayview Drive, Killiney, Co. Dublin. 01 2823584

H.A.D.D.
Hyperactive/Attention Deficit Disorder
Carmichael Centre, Nth Brunswick St. Dublin. 01 8748349

ASPIRE
Asperger's Syndrome Association of Ireland
Carmichael Centre, Nth Brunswick St. Dublin. 01 8780027

The Dyspraxia Association
54 Frankfurt Avenue, Rathgar, Dublin. 01 2957125

A.C.L.D.
Association for Children and Adults with Learning Disabilities
(incorporating the Dyslexia Association of Ireland)
1 Suffolk St. Dublin. 01 6790276

National Children's Resource Centre
Barnarados, Christchurch Square, Dublin. 01 4530355

National Association for Parent Support
Capoley, Portlaoise, Laois. 0502 20598

The School of Clinical Speech & Language Studies
Trinity College, Dublin. 01 6081496

Irish Deaf Society
Carmichael House, North Brunswick St. 01 8725748

The Minister of Education and Science
Marlboro St. Dublin. 01 8734700

Department of Education and Science
Special Education Section, The Prinicpal Office,
Athlone, Westmeath. 0506 21363

Department of Education and Science
Examinations Branch, Athlone, Westmeath. 0902 74621

National Parents Council – Primary
12 Marlborough Court, Dublin. 01 6789980

The National Educational Psychological Service
Frederick Court, Dublin. 01 8892700

Irish Autism Alliance
23 Summerfield Meadows, Blanchardstown. 087 9185002

SPECTRUM
This is an umbrella organisation which can be reached by telephone at (01) 874 8349. SPECTRUM is a support group that is composed of the following:

Dyslexia Association of Ireland (DAI)
Brainwave – The Irish Epilepsy Association
ASPIRE – Asperger's Syndrome Association of Ireland
HADD Family Support Group
Irish Association for Gifted Children
Dyslexia Awareness (formerly CHILD, Limerick)

The National Disability Authority
25 Clyde Road, Dublin 4. (01) 6080400

National Council for Special Education
1-2 Mill Street, Trim, Co. Meath. 046 9486400

Down Syndrome Ireland
30 Mary St., Dublin 1. 01 873.0999 1890 374 374

Tourette Syndrome Association of Ireland
Contact: Una Finucane
29 Granville Road, Dun Laoghaire, Dublin, Ireland.
TEL: (353) 1 623 0500 FAX: (353) 1 623 0500

Useful Websites

There are literally hundreds (perhaps thousands) of websites on the Internet for every sort of special education or medical condition. The reader is cautioned to be careful in accessing web information as some sites offer rather spurious strategies that have not withstood the test of scientific scrutiny. Also be careful to be aware of whether or not you are accessing an American website that relates to "Learning Disabilities". Remember – in America the term Learning Disability most often refers to the Specific Learning Disabilities such as Dyslexia. In America the term "Mental Retardation" is used to describe children we refer to as having "General Learning Disabilities". This is a critically important difference that can easily lead to confusion and result in using strategies that are inappropriate for a child.

The following are some that are of proven value to anyone interested in special education issues:

National Council for Special Education
http://www.ncse.ie/

American Foundation for the Blind
http://www.afb.org

Learning Disabilities Online
www.ldonline.org

American Hyperlexia Organisation
http://www.hyperlexia.org

Children and Adults with Attention Deficit Hyperactivity Disorder (CHADD)
http://www.chadd.org

The International Dyslexia Association
http://www.interdys.org

Scoilnet
http://www.scoilnet.ie/Scoilnet

British Dyslexia Association
http://www.bda-dyslexia.org.uk

Down's Syndrome Association-UK
http://www.downs-syndrome.org.uk

Down Syndrome Ireland
http://www.downsyndrome.ie

Tourette's Syndrome Association
http://www.tas-usa.org

Educational Resources Information Centre
www.eric.ed.gov/

Asperger's Syndrome Coalition of the US
http://www.asperger.org

Autism Society of America
http://www.autism-society.org

National Autistic Society-UK
http://www.nas.org.uk

Special Education Support Service
http://www.spss.ie

Irish Association for Teachers in Special Education (IATSE)
http://www.iatseireland.com

The State Examinations Commission
http://www.examinations.ie

Department of Education and Science
http://www.education.ie

Genetic and Rare Conditions Website
http://www.kumc.edu/gec/support/

Irish Learning Support Association
http://www.ilsa.ie

National Council on Curriculum and Assessment
http://www.ncca.ie/

Irish National Teachers Organisation
http://www.into.ie

National Centre for Technology in Education
http://www.ncte.ie

National Technology and Special Needs Advisor
Email: ajackson@crc.ie

APPENDIX THREE

Private Psychological Assessments

Primary schools that do not have access to NEPS may commission private assessments under a scheme administered by NEPS. Each school may commission a number of assessments (approximately one assessment for every fifty students). Under this scheme, individual psychological assessments may only be administered by psychologists whose names appear on a panel compiled by NEPS. The reader is advised to review the websites of NEPS and the DES for updated information on this panel.

The following table from the NEPS website indicates how many private assessments a school without access to the NEPS service may apply for.

Schools without a NEPS service	
Number of pupils	Number of assessments
1 – 50	1
51 – 100	2
101 – 150	3
151 – 200	4
201 – 250	5
251 – 300	6
301 – 350	7
351 – 400	8
401 – 500	10
501 +	12
continuing with an additional 2 assessments per 100 students	

In certain circumstances, some NEPS supervisors have the discretion to request private assessments in schools which *have* a NEPS service available.

NEPS Active Panel

For a list of the psychologists currently registered on the panel approved by NEPS to complete private assessments, please go to the following website: http://www.education.ie/servlet/blobservlet/neps_active_panel.xls

For more information, contact:
National Educational Psychological Service, 24-27 North Frederick Street, Dublin 1. Tel: 01-889 2700. Email: neps@education.gov.ie

APPENDIX FOUR

School Transport

Children with special needs enrolled in a special class in a primary school, or in a special school, may qualify for school transport. The school principal applies for transport through the school inspector. In the case of children enrolled in special schools for the Visually or Hearing Impaired, the application is made directly to the DES by the school principal. If a child is attending a residential school, transportation can be arranged on weekends in some cases. There is no guarantee that transport will be made available. The DES investigates each case on its own merits.

If transport cannot be provided, a special grant is made available to parents to assist with the cost of private transport. A transport grant is paid under the following conditions according to the DES document entitled "School Transport for children with Special Needs)", available on the DES website:

- If the child has to be brought to a specific pick-up point to meet a special transport service
- If the child requires a supervised transport service and it is not possible to provide this service
- If there is no special transport service available for a special needs child travelling home from a residential school at the weekend.

The rate at which the grant is paid depends on mileage and the attendance at school of the child. The grant rates quoted in the DES statement are as follows (as of September 1999):

Distance:	€ Amount Per Annum:
0 – 0.5 miles	€140
0.6 – 1.0	€195
1.1 – 1.5	€255
1.6 – 1.9	€305
2.0 – 2.5	€370
2.6 – 3.0	€420
3.1 – 3.5	€470
3.6 – 3.9	€535
4.0 – 4.5	€585
4.6 – 5.0	€650
5.1 – 5.5	€700
5.6 – 5.9	€750
6.0 – 6.5	€815
6.6 – 7.0	€865
7.1 – 7.5	€930
7.6 – 7.9	€980
8.0 – 8.5	€1030
8.6 – 9.0	€1095
9.1 – 9.5	€1145
9.6 – 9.9	€1210
10.0 – 15	€1400
15.1 – 25	€1755
25.1 +	€2100

An appeals procedure is in place related to these grants and appeals can be granted for various conditions of 'hardship', which may include financial difficulties, family circumstances, availability of the family car at certain times, etc.

For children in special schools, or special classes in mainstream school, the DES has limited funding available to supply a harness for the child if necessary. Additionally, each year a sum of money is made available to provide an escort for the child.

Special transport is only made available for children in special schools or special classes in mainstream schools, and terminates once the child returns to the mainstream. If a child with a special need is enrolled in a mainstream class the parents can apply directly to the School Transport Section, Department of Education & Science, Tullamore, Co. Offaly. Tel: 0506-24351/2/3/4/5/6.

Glossary

Ability Grouping – Placing students of similar ability in the same class or group for purposes of instruction

Activities of Daily Living (ADL) – Everyday skills the person needs to learn to function: eating, dressing, bathing, hygiene skills, communication skills

Adaptations – Strategies that support students in achieving the learning outcomes of the curriculum. Adaptations may include adjustments to assessment, instruction, environment or resources

Adaptive Development – How a child compares to other children the same age, as regards such things as motor development, speech and language skills, daily living skills (ADL's) etc.

Adaptive Physical Education – Physical education programmes modified to meet the needs of special education students

Aptitude – Undeveloped potential or ability

Asperger's Syndrome – Sometimes referred to as 'High Functioning Autism'; a condition similar to autism but without the significant delay in speech development in early childhood; usually not as severe as autism and not associated with General Learning Disability; it is a neurobiological condition

Assessment – A systemic process for gathering information from many sources in order to make appropriate educational decisions for a student. It is a progressive process identifying the student's strengths and needs. It results in the design and implementation of specific educational strategies. Assessments are conducted on a continuum and can be both formal and informal in nature

ADHD (Attention Deficit Hyperactivity Disorder) (ADD) – Student exhibits poor attention, distractibility, impulsivity, and hyperactivity

Advocate – A person who helps take action for someone else who is not able to

Age Appropriate – Within the child's chronological age

Age Norms – The average performance of an individual in various age groups

Annual Goals – Yearly goals documented in the Individualised Education Plan

Anti-social Behaviour – Behaviour that does not meet the minimal standards of acceptability for the environment in which the person is embedded

APGAR Score – Score given to a newborn to identify infants at risk – colouring (appearance), heart rate (pulse), muscle tone (activity), breathing (respiration) and response to stimuli (grimace) are assessed

Articulation Disorders – Difficulty with the production of speech sounds

Asynchronous Development – Differing rates for physical, cognitive and emotional development (also known as 'dyssynchronous development'). For example, a gifted child may be chronologically 13 years old, intellectually 18, emotionally 8, and physically 11. The discrepancies are greatest for everyone at the chronological age of about 13, but the extremes displayed by gifted children have led some experts to define giftedness itself as asynchronous development. If you tell a gifted child to "Act your age!" he/she could legitimately respond: "Which one?"

Autism/Autistic Spectrum Disorders – A severe childhood disorder involving extreme social isolation, difficulty communicating, unusual behaviour, echolalia (repetition of words), limited mental abilities and delayed development

Brain Lateralisation – Specialisation of the brain hemispheres. In right-handed people, the right brain hemisphere is more involved with spatial relations, imagery, and non-verbal, non-sequential processing, while the left-brain hemisphere is more involved in verbal and sequential processing

Characteristics of the Gifted – The following characteristics are common but not universal:
- Shows superior abilities to reason, generalise or problem solve
- Shows persistent intellectual curiosity
- Has a wide range of interests; develops one or more interests to considerable depth
- Produces superior written work or has a large vocabulary
- Reads avidly
- Learns quickly and retains what is learned
- Grasps mathematical or scientific concepts readily
- Shows creative ability or imaginative expression in the arts
- Sustains concentration for lengthy periods on topics or activities of interest
- Sets high standards for self
- Shows initiative, originality, or flexibility in thinking; considers problems from a number of viewpoints
- Observes keenly and is responsive to new ideas.
- Shows social poise or an ability to communicate with adults in a mature way
- Enjoys intellectual challenge; shows an alert and subtle sense of humour

These characteristics can lead to conflicts in the regular classroom, as the gifted child may:
- Get bored with routine tasks
- Resist changing away from interesting topics or activities
- Be overly critical of self and others, impatient with failure, perfectionistic
- Disagree vocally with others, argue with teachers

Glossary

- Make jokes or puns at times adults consider inappropriate
- Be so emotionally sensitive and empathetic that adults consider it over-reaction; may get angry or cry when things go wrong or seem unfair
- Ignore details, turn in messy work
- Reject authority, be non-conforming, stubborn
- Dominate or withdraw in cooperative learning situations
- Be highly sensitive to environmental stimuli such as lights or noises

These reactions of gifted students to the regular education environment are only within the context of an understanding of the gifted. Without that understanding, they may be used to label the student as or Emotionally Behaviourally Disturbed

Cause and Effect – The ability to understand that a specific action can make something happen

Clinical Psychologist – a psychologist who specialises in the emotional, intellectual, emotional, behavioural and adjustment problems of children who may have mental health difficulties such as depression, ADHD, bi-polar disorder.

Collaboration – An interactive process where a number of people with particular expertise come together as equals to generate an appropriate programme or process or to find solutions to problems. Collaboration is an important skill when children are engaged in group work. Children must be taught how to collaborate before they can successfully engage in group work

Cognitive – A cognitive disorder is a difficulty in learning in the areas of reasoning, comprehension and judgment

Congenital – Present at birth

Cooperative Learning – Students working in small groups, where often the same assessment result is given to all. Students motivated by achievement may dominate the group and do all the work so their own grades don't suffer, and may simply withdraw or refuse to participate. Cooperative learning groups with students of *similar* ability, with complementary skills, tend to work most smoothly

Criterion Referenced Test – Child is evaluated according to own performance, not in comparison to others

Curriculum Differentiation – Changing or adapting the methods of teaching and instruction, the resources, and the classroom environment to meet the needs of all learners

Developmental Disability – A condition that prevents a child from developing normally and often results in a general learning disability or autism

Diversity – The ways in which individuals differ from each other. Differences may be visible (for example, race, gender, age); others are less visible (for example, culture, ancestry, language, religious beliefs, sexual orientation, ability, socio-economic background)

Down Syndrome – Down syndrome is a chromosomal disorder characterised by the presence of an extra #21 chromosome. Instead of having 46 chromosomes in each of his/her cells, a person with Down syndrome has 47. The extra chromosome may be derived from either the egg or the sperm

Drumcondra Primary Reading Test – a test of children's reading achievement, which compares each child's score to the national average

Dyscalculia – A learning disability in which a child is unable to do maths problems

Dysgraphia – A learning disability that impairs the child's ability to write

Dyslexia – Learning disability affecting reading ability. Persons with dyslexia may have difficulty remembering, recognising, and/or reversing written letters, numbers, and words, might read backwards, and have poor handwriting

Dyspraxia – Impairment or immaturity in the organisation of movement

Education Program – An organised set of learning activities designed to enable students to develop to their potential and acquire the knowledge, skills and attitudes they need to contribute to a healthy society

Educational Psychologist – a psychologist who specialises in the learning, adjustment, development, emotional and social aspects of the lives of children in school

Emotional Disturbance and/or Behavioural Problems – Having a long-standing, severe condition that adversely affects personal adjustment, social relationships and learning

Extrinsic Motivation – 'Reinforcers', rewards or incentives used by one person to bring about desired behaviour in another person; they often take the form of little gifts, stickers, verbal and non-verbal praise

Fine Motor – Hand and finger small muscle movement, coordination and control

Formal Assessment – Includes psycho-educational assessment by a psychologist to determine academic skill development, intellectual functioning, strengths and weaknesses in learning processes, and social and adaptive skills. Formal and informal assessments help teachers and parents to better understand the nature of the special need and how to choose appropriate educational strategies for the child

Froebelian Method – an educational philosophy and practice based on the theories of Friedrich Froebel (1782-1852) that emphasises the importance of play, the need for spontaneity, self-activity, creativity, co-operation, and places an emphasis on the use of concrete materials in a joyful, harmonious classroom atmosphere

General Learning Disabilities – Conditions resulting in significantly sub-average intelligence, ranging in severity from Borderline, Mild, Moderate, Severe and Profound levels

Gentle Teaching – a method of teaching children with special needs that was devised in the Netherlands and focuses on the child's need to feel safe, engaged, unconditionally loved and to feel loving towards caregivers

Gifted/Exceptional Ability – see "Characteristics of the Gifted" above

Gross Motor – Coordinated movements of all body parts including the arms and legs

Guardian *ad litem* – A person appointed by the court to represent the rights and interests of a child

Hearing Impairment – A hearing disability ranging from mild to profoundly severe

Hyperactivity – excessive motor activity or restlessness

In Loco Parentis – To assume the duties and responsibilities of a parent and to do what a "reasonable" parent would do. Teachers act in *loco parentis* to the children in their classroom

Inclusion – The value system which holds that all students are entitled to equitable access to learning, achievement and the pursuit of excellence in all aspects of their education. The practice of inclusion transcends the idea of physical location and incorporates basic values that promote participation, friendship and interaction. Disabled children receive services in their home school and are placed in the same classroom with non-handicapped children

Individual Education Plans (IEPs) – Written records that document the individualised planning processes for students with special educational needs. Individualised planning is a continuous and integrated process of instruction, assessment, evaluation, decision-making and reporting

Informal Assessment – Includes student observations, file reviews, team consultation, interviews, and academic or behavioural testing to determine skill development, strengths, and weaknesses. Formal and informal assessments help teachers and parents to better understand the nature of the special need and how to choose appropriate educational strategies for the child

Integration – The practice of placing students with special needs in educational settings with peers who do not have special needs

Intelligence Quotient (IQ) – relates to the common understanding of human intelligence; it is measured by the administration of a test on a one-to-one basis; IQ is a concept that has come under increased scrutiny in recent time; a person's IQ is a good predictor of how successful they will be in school but a poor predictor of how successful they will be in other areas of life

Interventions – The putting into practice of plans, strategies and supports to facilitate learning and to address the special needs of students

Intrinsic Motivation – The desire that comes from within the individual to satisfy natural needs and interests, which includes a desire to understand and make sense of the world

Labelling Theory – The proposition that labels placed on a person may lead him/her to act the role associated with the label, whether or not it was initially accurate. When others know a label, they may interpret the labelled person's behaviour as abnormal, whether it is or not. This changes their actions toward the labelled person so that their interactions reinforce the label

Learning Difficulty – different from a special education condition, a learning difficulty may be the result of mild developmental delays or mild problems adjusting to school; learning difficulties are not as serious as special education conditions, do not require the same level of support or structure and often disappear after mild intervention or assistance

Levels of Giftedness – According to measurements, the following labels are generally accepted:
- Bright – 115 and above
- Gifted – 130 and above
- Highly gifted – 145 and above
- Exceptionally gifted – 160 and above
- Profoundly gifted – 175 and above

LRE - Least Restrictive Environment – A child should be educated in the least restrictive environment for his or her disability and which meets his or her needs

Mainstreaming – Some or all of the child's day is spent in a regular classroom

Mean – Arithmetical average

Median – A measure of central tendency, where half the scores are above and half below

Mental age – The level of intellectual functioning based on the average for children of the same chronological age

Mental Retardation – the terminology used in America to identify and label people with what in Ireland are known as 'General Learning Disabilities'; the IQ ranges for Mental Retardation differ from those associated with the GLD's as follows:

Mental Retardation		*General Learning Disabilities*	
Borderline	**70-79**	**Borderline**	**70-79**
Mild	55-69	Mild	50-69
Moderate	40-54	Moderate	35-49
Severe	25-39	Severe/Profound	below 35
Profound	under 24		

Micra-T Reading Attainment Test – The MICRA-T is Ireland's leading reading test. The principle purpose of the test is to provide Irish primary school teachers with accurate information on the reading levels of pupils in their classes. In particular, the tests enable teachers to compare the reading performances of their pupils with reading standards nationally

Mixed Ability Grouping – grouping children of differing achievement levels together in the same class or learning group within a class

Mode – The most frequent score

Modifications – Changes to learning outcomes that result in students' learning outcomes being substantially different from those of the provincial curriculum. Modifications are specifically selected to address a student's special needs

Montessori Method – An educational philosophy based on the ideas of Italian physician/educator Maria Montessori (1870-1952). Although originally developed with students labelled "mentally defective" her tremendous successes led her approach to be widely embraced, especially in upper class pre- and primary schools worldwide. Montessori saw students' learning as the result of innately self-motivated activity. The teacher's job, then, is to supervise and guide rather than transmit knowledge

Motor Development – The ability to move effectively within the environment

Glossary

Multidisciplinary Team – a team of specialists such as a speech and language pathologist, psychologist, occupational therapist, used to help determine the student's needs

Multiple Disabilities – Two or more special education conditions occurring in the same person at the same time

Muscle Tone – The amount of tension in the muscles when at rest; muscle tone in children with cerebral palsy and other physical disabilities is often tight and requires physiotherapy

Neonatal – Period between onset of labour and several months after birth

Neurobiological Condition – Conditions resulting in special education need which arise from faulty electrical-chemical functions within the brain

Norm – (1) In sociology, a culturally relative guideline for social behaviour. (2) In testing, a statistical measure of central tendency, as a mean, median, or mode

Norm Referenced Tests – A child's performance is compared to others the same age

Occupational Therapist – A therapist that focuses on daily living skills, sensory integration, and fine motor skills

Percentile – a score which reflects how a child compares with his or her peers; based on a mathematical standing out of one hundred; e.g. a percentile score of seventy-five means that twenty-five children out of one-hundred will have a score equal to or greater than the child's score

Perinatal – The period of time at or immediately following birth

Physical Disabilities – Having a condition that impairs the normal development of muscle activities (e.g., having spina bifida, cerebral palsy, or loss of limbs); a category of special education

Physiotherapist – Provides evaluation and treatment of physical disabilities to help the person improve the use of bones, muscles, joints, and nerves through exercise and massage

Placement – The level, grade, classroom or setting that is most suitable for the student

Postnatal – Period of time after birth

Pragmatic Language – that aspect of language usage and understanding that relates to humour, sarcasm, understanding of idioms, ability to take turns in conversation, etc.

Prenatal – Period of time before birth

Pro-social Behaviour – Behaviour that acceptably meets the standards of the social environment in which one is embedded

Rating Scale – a form that can be completed by an adult to help determine if a child's behaviour, learning or development is significantly different from that of other children the same age

Receptive Language – The understanding of spoken and written communication as well as gestures

Reliability – The accuracy and repeatability of a measurement

School-Based Team – An ongoing team of school-based personnel which functions as a problem-solving group to assist classroom teachers in developing and implementing instructional and management strategies and to coordinate support resources for students with special needs within the school

Sigma-T Mathematics Attainment Test – a test of mathematics achievement, like the Micra-T Reading Attainment Test, linked with the Revised Primary School curriculum

Specific Learning Disability – A child with average or above average potential has difficulty learning in one or more areas (such as reading or maths) and exhibits a severe discrepancy between their ability and achievement

Specific Speech and Language Disorder – Disability involving impaired receptive or expressive language skills, stammering or serious articulation problems

Standardised Test – A test taken by many students under identical conditions which allows results to be compared statistically to a standard, such as a norm

Students with Special Needs – Students who have disabilities of an intellectual, physical, sensory, emotional or behavioural nature, who have a learning disability or have exceptional gifts or talents

Tactile Defensiveness – Child overreacts or avoids any kind of touch. It is a symptom associated with autism or more severe ADHD

Tourette's Syndrome – a neurological disorder characterised by tics or motor movements and often occurring with ADHD/Specific Learning Disabilities

Tracking – Full-time, often-permanent assignment to achievement groups, i.e. children with the same level of academic standing in a subject in the same group

Transition – The passage of students from one environment to another at key points in their development, from childhood to adulthood, for example: into infant classes or from primary school to secondary school or from secondary to third level, or from third level to workplace. For some children with special education needs transition can also be from secondary level to the workplace, sheltered living centre, etc. Schools should provide transition services to assist a child with disabilities to successfully access the adult world, through work experiences and/or through postsecondary options and related. Transition services must be individually tailored to the child's needs and skills

Underachievement – A significant difference between ability and performance. A gifted underachiever is often defined as having superior intelligence, yet working below grade level. Underachievement is sometimes differentiated from by including a psychological factor of perceived inability to succeed academically. Some underachievers may withdraw, others may become disruptive. Factors that can contribute to underachievement include:
- Lack of respect for the individual
- An overly competitive environment
- Inflexible and rigid structure
- Stress on external evaluation and criticism
- Authoritarian control
- Unrewarding curriculum
- Family conflicts, such as divorce

Validity – (1) In testing or - A measurement's ability to measure what it purports to measure

Visual Impairment – limited vision that, even with correction, adversely affects a child's educational performance; a category of special education for children with such a condition

William's Syndrome – Williams syndrome is a rare genetic condition (estimated to occur in 1/20,000 births) which causes medical and developmental problems. Williams syndrome was first recognized as a distinct entity in 1961. It is present at birth, and affects males and females equally. It can occur in all ethnic groups and has been identified in countries throughout the world. This syndrome is often associated with a General Learning Disability and academic difficulties

Index